Einstein Plus Two

By

Petr Beckmann
Professor Emeritus of Electrical Engineering,
University of Colorado
Fellow,
Institute of Electrical and Electronic Engineers

THE GOLEM PRESS
Boulder, Colorado
1987

Library of Congress Catalog Card No.: 85-82516
ISBN 0-911762-39-6

THE GOLEM PRESS
Box 1342
Boulder, Colorado 80306

To the two great physicists of our time,

Edward Teller

and

Andrei Dmitriyevich Sakharov

Preface

When I run, I feel a wind; but not one that will make a windmill turn.

As long as an observer is at rest on the ground, it does not matter whether the velocity of the wind is referred to the observer or the windmill. A physicist who *falsely* assumes that the effect-producing velocity (that makes the windmill turn) is that with respect to the observer, but *correctly* applies the relativity principle, will expect the windmill to turn when he is running. The experimental evidence will contradict his expectation, and he can then either abandon his false premise, or he can so distort space and time that the observer's motion produces two exactly equal and opposite forces on the windmill, keeping the mill motionless as observed. The Einstein theory, in effect, takes the latter road; but I believe the laws of physics, including the relativity principle, must hold regardless of any observer, who should do nothing but observe.

An electric or magnetic field will accelerate an electron. Its magnetic field will therefore increase, which causes the induced electric field to decelerate it. That will decrease the magnetic field and the induced electric field will accelerate the electron again. The resulting oscillations are derived from the Maxwell equations in Part Two of this book. They explain the quantization of electron orbits, the de Broglie relation and the Schrödinger equation simply and without further assumptions.

The natural frequency of these oscillations depends on the velocity of the electron; but the velocity with respect to what? The velocity that will make the Lorentz force and the Maxwell equations valid, claims the Einstein theory, is the velocity with respect to the observer. But if so, does the electron oscillate for me because I am moving past it, but not for you because it lies still in your rest frame? To answer yes is to kill the relativity principle.

As I will attempt to show, the velocity that makes the Maxwell-Lorentz electrodynamics valid is that of charges with respect to the local fields they traverse. That squares with the experimental evidence in electromagnetics and optics, and it leads to the derivation of two phenomena for which no explanation other than *ad hoc* postulates has hitherto been available: the quantization of electron orbits and in the realm of gravity, the Titius series.

Why, then, has the Einstein theory celebrated an uninterrupted series of brilliant successes for more than 80 years?

Because in all past experiments the observing instruments have always been nailed to the local field, so that they could not reveal whether the observed effect was associated with an observer-referred or a field-referred velocity. The technology for testing that difference may not be available for some time.

But if it is field-referred velocities that are the *effect-producing* ones, then the Maxwell equations automatically become invariant to the Galileian transformation; the undisputed fact that the Lorentz force and the Maxwell equations with *observer*-referred velocities are Lorentz-invariant is one that becomes both trivial and irrelevant.

I am not so naive as to think that the first attempt to move the entire Einstein theory *en bloc* onto classical ground will turn out to be perfectly correct. What I do hope is that the approach will provide a stimulus for the return of physics from description to comprehension. Attempting to redefine the ultimate foundation pillars of physics, space and time, from what they have been understood to mean through the ages is to move the entire building from its well-established and clearly visible foundations into a domain of unreal acrobatics where the observer becomes more important than the nature he is supposed to observe, where space and time become toys in abstract mathematical formalisms, and where, to quote a recent paper on modern approaches to gravitation theory, "the distinctions between future and past become blurred."

This book is for those who do not wish to blur such distinctions ("He will commit posthumous suicide yesterday"?). It is for those who seek to understand rather than merely to describe; for those who will accept the Einstein theory as a brilliant, powerful and productive equivalence, but not as a physical reality.

It is for those who are prepared to sacrifice a lifetime's investment in learning; and perhaps more importantly, for the young students who have not yet made such an investment.

Boulder, Colorado *P.B.*
1983-1987

Contents

Introduction:
Truth and Equivalence

Introduction:
Truth and Equivalence

Truth, some say, is what agrees with experiment.

Necessary, but not sufficient: fata morganas can be photographed, and all astronomic measurements on earth record the same position of a star that may not exist: next week's observations of the star's light may bring the news that it blew up in the 14th century. The mirror image of a candle behaves as if it were emitting light, and a body immersed in water behaves as if it had lost mass.

To develop a workable guideline for what is true and what is equivalent, consider first some uncontroversial cases of equivalence.

A good example is provided by the ionospheric equivalence theorems (there are two, but for our purposes they can be merged into one). When radio waves are returned by the ionized layers in the upper atmosphere, they are not reflected by them like a tennis ball is bounced off a wall. Even using the geometric optics simplification, a radio pulse travels with variable speed along a path similar to the one sketched in the figure below.

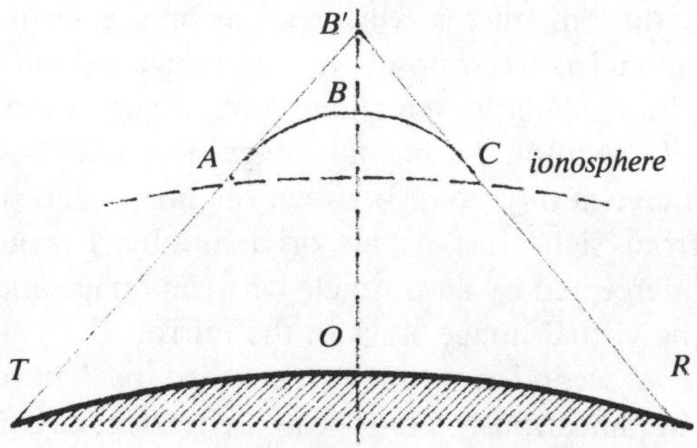

The ionospheric equivalence theorem (true and effective height). A radio wave pulse is slowed along the segment ABC in the ionosphere, but the transit time is the same as if it ran the path $AB'C$ with constant velocity c.

On entering the ionized layer at A, the pulse slows and the path curves (for reasons given in any textbook of ionospheric propagation) until at the point B it becomes horizontal and the pulse comes to a standstill – in the geometric optics approximation, anyway. The process then reverses itself symmetrically, and the pulse leaves the ionized layer with the velocity of light at the point C.

It is not a simple process, and the ionospheric equivalence theorems provide welcome relief: as proved in any textbook on ionospheric radio wave propagation, the time taken by the pulse to make it from transmitter T to receiver

R by the true, curved, slow path via B is exactly equal to the time that would be taken by a fictitious pulse traveling with constant free-space velocity from T to R via the straight sides of the triangle with apex at B'.

Thus the true height of the reflection point OB is replaced by the *effective height* (the actual term used in ionospheric research) OB' of the reflection point, which is the height that the pulse would reach in the same time if it propagated with the velocity of light throughout the trip. Since an ionospheric station, like any other radar, measures the time elapsed between transmission and reception, the two are equivalent. The real height is true, but involves bothersome calculations; the effective height is fictitious – a "just as if" equivalent height – but much simpler to use. (The two heights are related by a Volterra integral equation.)

Now here it is quite uncontroversial which of the two heights is true as a physical reality, and which is merely equivalent in producing the same effect on the measuring instrument. The obvious criterion for distinguishing between the two is that the effective height has limited validity: it will work when we measure the time for the echo to return, but not otherwise. A satellite measuring ionization directly will agree only with the real height of the layer, as will any other independent method.

Limited validity is, in fact, the first of my two proposed guidelines of how to separate truth from equivalence.

Consider two more examples of the limited-validity guideline.

A real image of an object is one whose points are sources of optical rays, just as they are on the original object. A real image is, for example, produced by an object located beyond the focal distance of a concave mirror. But the plane bathroom mirror will produce only a virtual image – it is *just as if* the rays emanated from points on the image behind the wall, but in reality they do not. Limited-validity guideline: a real image behaves optically like a real object under *all* conditions; a virtual image only under some. Intercept the rays from a real image at *any* point between the image and the eye, and the image will disappear from sight just as an original object would. But if the equivalent rays are intercepted by an obstacle between image and eye just behind the bathroom wall, the virtual image stays in the mirror.

A second example is provided by Thévenin's Theorem, which permits the simplification of complicated electrical circuits. It states that in any linear circuit the voltage between any two points, such as A and B on p.13, is the same as if it were caused by a single source in series with a single impedance (with values also given by the theorem). Truth and equivalence are sharply separated here by the limited-validity guideline: Let us assume that figure (a) represents a real circuit, and (b) is the equivalent circuit calculated by Thévenin's Theorem. Then in figure (b) what is to the right of A-B corresponds to voltages, currents and

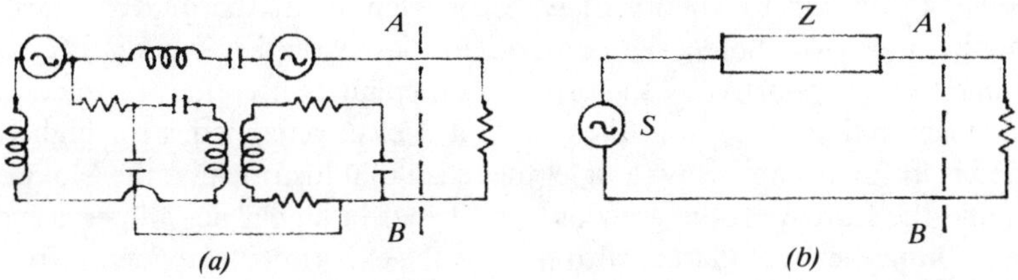

Thévenin's Theorem

circuit elements in the real world; but what is to the left of it is a "just as if" mathematical equivalence which is fictitious, notwithstanding the fact that such a circuit *could*, if we so desired, be very easily realized as a physical reality.

Let us now apply the limited-validity guideline to the Einstein Theory (there is a good reason why I am reluctant to call it the Theory "of Relativity"). Is it limited or universally valid?

It is certainly universally valid in its claims, and there is no experimental evidence to contradict it. However, such evidence can be obtained only when sources of light or elementary particles move with a velocity comparable with the velocity of light, and this, at present, restricts the *verified* results to a surprisingly narrow field: a handful of optical experiments (which are also supported by an alternative hypothesis), and electromagnetics – and please hold back your protest until I fully explain what I mean.

First, I have singled out the optical experiments because they make no use of the electromagnetic nature of light. They use light simply as something that has the capacity to interfere and that travels from here to there with velocity c.

The rest of the acceptable evidence virtually always relies directly or indirectly on electromagnetic theory, as will be shown in Part One. In particular, the velocity of elementary particles is rarely measured directly (as, say, the ratio of distance covered to time elapsed), but is usually *inferred* from the directly measured voltage and the Lorentz force, which is *assumed* to remain valid at high velocities, which are *defined* to be velocities with respect to the observer. Similarly, the decrease in the *ratio* of charge to mass of elementary particles at high velocities is always attributed to the increase in inertial mass, for the invariance of electric charge has simply been *postulated*. More examples will be given in Part One, where these points will be discussed more fully.

This faith in the extrapolated validity of our presently accepted electromagnetics at high velocities makes the Einstein theory very different from other universal principles in physics. The law of the conservation of energy, for example, has been verified in all branches of physics and beyond — biology and chemistry, for example. If the kinetic theory of gases or even all of thermodynamics were to collapse tomorrow, the energy conservation law would not budge, for it would continue to be supported by the orbits of the planets, the tides of the ocean, and the nitrogen-fixing bacteria in the soil. But if electromagnetics for high velocities were to be refuted tomorrow (and let me recall that historically, the Maxwell equations and the Lorentz force grew out of a belief in an elastic, all-pervasive ether), the first thing they would take with them is the experimental evidence for the Einstein theory.

Note that I am not complaining about the *amount* of supportive evidence for the Einstein theory; only a crank (and there seem to be plenty) would go to war against Einstein on that account. What I am complaining about is the narrow field from which this plentiful evidence is gleaned.

No length contraction has ever been shown on a well-defined, charged or uncharged body with well-defined dimensions and a velocity measured by several independent methods, if not directly; no time dilation experiment has ever provided proof that the changed rate of the clock is only perceived by the moving observer and has not taken place in the clock itself.

The Einstein theory has never proved its two *tacit* postulates: that the Maxwell-Lorentz electrodynamics, remain valid at high *observer-referred* velocities; and that the motion of matter through a force field does not inherently — independently of any observer — change its own force field.

Yet this very objection also shows that the limited-validity criterion is not (or not yet) usable on the Einstein theory. Without an experimental refutation of the theory, we do not know whether its limited validity is inherent, as it is in a virtual image, or whether it is merely due to our technological limitations in being unable to impart a sufficiently high velocity to anything but elementary particles.

Let us then examine another possibility for distinguishing between truth and equivalence when the difference cannot be established by full vs. limited experimental confirmation. For this purpose I have thought up the Grandiose Theory of the Railroad Track.

The rails of a railroad track appear to converge as they recede into the distance, as we have all seen with our own eyes; yet we all know that in reality they are reasonably parallel. The reason why nobody considers that a paradox, I suppose, is that we have learned from childhood to trust our mind and experience when our eyes deceive us — for railroad tracks if not for TV documentaries.

The explanation is "perspective" — the way in which images are projected onto the retina or onto the camera's focal plane. It is not terribly complicated, but it is not the simplest thing in the world, either: most of us would rather pay for ready-to-use perspective software than go through the chore of writing it ourselves.

But my Grandiose Theory of the Railroad Track offers an alternative explanation: lengths shrink with distance from the observer.

Now you and I know that this is an absurdity, but imagine some Martian Mole, who is intelligent, logical and erudite, but has no means of remote sensing. Suppose he visits us and wants to know why humans perceive a railroad track as converging, and is given the two theories: perspective and distance-shrink. "I use Ockham' s razor," he might say, "and I buy the shrinkage theory. Of the two, it is by far the simpler."

Don't try to use measuring rods; they contract as they are carried away from you along the track; and don't go there with the measuring rod yourself, because the track will shrink behind you.

A closed loop with an interferometer? No: the wavelength shrinks with distance from the observer — that's why railroad tracks are notorious for the absence of fringe shifts.

But if the wavelength changes without producing a Doppler effect, the frequency of the light must have changed, you say. Of course it has; have you never heard of time dilation?

You install a second beamsplitter and interferometer (plus observer) at the far end of the loop, proving that the distance between the rails is the same at both ends of the track at the same time.

But you have proved no such thing. The wavelength is shortened away from the observer: it shrinks for one *this* way, and for the other *that* way, and each observer observes, from his own point of view, the same outcome of a different process. That's what modern physics understands by "relativity;" and whatever measuring instrument we may use is subject to the same perversion as the quantity it attempts to measure.

Now suppose the theory could not be disproved experimentally; how would we know it is absurd?

To some extent, of course, the flaw in the Grandiose Railroad Theory lies in the fact that, like the Einstein theory, it is not tied to nature itself, but to the observer or instrument that measures it. If I had tied the contracting distances to Grand Central Station, you would not need an interferometer to disprove it; you could go uptown and jump across the tracks.

However, observer-dependence in itself need not be flawed. Velocity is observer-dependent; it has no meaning unless we specify with respect to what standard of rest we measure it. Some functions of velocity — such as the Doppler effect — must necessarily be observer-dependent, too.

In the *Dialogues on Two World Systems*, Salviati, fronting for Galileo, took great pains to persuade Simplicio, representing Aristotle and the Church, that the path of a stone dropped from the mast of a moving ship would appear oblique to a stationary observer on shore, though it would hit the deck at the same distance from the mast as when it was dropped ([Galileo 1630], pp. 142-144 of the English translation). There is no flaw *per se* in certain quantities being observer-dependent.

But not *everything* is observer-dependent. Note that in Galileo's example the velocity vector is observer-dependent, but the distance (from the mast) is not. Surely space and time must be the same for all objects dwelling in them if any consistency is to be preserved; and our measuring standards, if they are to be standards, must not be subject to the fluctuations of the quantities of which they are supposed to be standards.

Both the Einstein and the railroad track theories break that rule, and they do so in the particularly critical case of space and time, which are something special in that together with mass, they are used to define velocity, momentum, acceleration, force, and progressively higher concepts. But space and time themselves cannot be defined; if they could, any non-circular definition would have to involve a more primitive concept still. When a philosopher says that time is "that which flows from future into the past" he is using descriptive lyrics, not a one-to-one mathematical definition.

Assuming that the railroad track theory could not be disproved by direct experiment, it could be recognized as (at best) an equivalence by its tampering with the fundamental, and hence undefinable concepts on which everything else is built; and this tampering with the primitive fundaments is what I propose as a second guideline for discerning truth from equivalence. Mathematics is perfectly free and unfettered by experimental observation to define its axioms from which it deduces their consequences; physics, if it is to understand the real world, must build on the two primitive and undefinable pillars. It must not tamper with them in order to accommodate higher concepts. It must not redefine the undefinable; more particularly, it must not make the primitive pillars observer-dependent.

Note that this proposal has nothing whatever to do with "absolute" space or "absolute" time. We are still free to choose the origin of our coordinate system in both space and time where we please, for there is no evidence of any system being more privileged (though it may often be much more convenient) than any other. And most certainly we need not give up the Principle of Relativity.

The Einstein theory, then, may not turn out as general as experiments relying on presently accepted electromagnetics make it appear; and it defines the undefinable primitives space and time via the higher-order concept of velocity, arguably making the definition circular, and certainly making the two primitives observer-dependent.

But there is a third point that makes it highly suspicious: One of its two postulates may be *inherently* irrefutable.

A theory may be irrefutable because it is true; or it may be irrefutable because it is inherently protected against refutation, even though it may be false. A crude example of a theory that is close to irrefutable, but patently untrue, would be the claim that the earth has a second moon, made of a material that becomes perfectly transparent when illuminated.

The Einstein theory rests on two postulates. The first is the Principle of Relativity, known for more than three centuries, with which few reasonable men will quarrel. But the other, known as the Second Postulate, postulates a constant velocity of light *independent of the state of motion of the emitting source* (and therefore, by the relativity principle, also independent of the state of motion of the receiver).

With respect to what? In the Einstein theory, with respect to the observer: if two observers move with different velocities with respect to the same source, each measures the same velocity of its light. This is not only sharply different from what we are used to with low velocities, but plays havoc with space, time and simultaneity. The usual explanation for this bizarre postulate is that there is no reason why we should expect high velocities to add in a manner linearly extrapolated from our experience with low velocities.

But the Second Postulate violates a lot more than unimaginative thinking; indeed, it violates a lot more even than the time-honored concepts of space and time.

Imagine that the Second Postulate were valid, on some planet in a distant galaxy, not for light, but for water squirted from a fountain in periodic pulses acting as time signals. No matter whether you stood still, ran with the water or against it, you would always measure the same velocity of the water with respect to yourself.

Would this have to be a planet where space and time are something quite different from what we are used to? Not at all: it would have to be no more than a planet on which nothing moves *faster* than the water squirted by the fountain (with standardized velocity and pulse frequency). You would then set c in the Lorentz transformation equal to the velocity of the water and proclaim it a universal constant — and the Lorentz transformation will do the rest, for it will so distort space and time that it will *force* the water postulate to be "true," i.e., agree with measurement.

All measurements would keep confirming the water postulate beautifully due to Einstein's theorem for the addition of velocities as long as only velocities slower than that of the water are used. Suppose, for example, that this imagined planet is inhabited by highly intelligent beings who are, in our vocabulary, deaf and blind, and the water squirted from the fountain in their National Bureau of Standards is the fastest thing they know. The theory would be much acclaimed, because it predicts everything correctly in spite of its bizarre water postulate.

But there is a flaw: the theory is revealed as incorrect one day when a scientist discovers the microphone and makes sound dectectable by his people's senses. He uses sound signals to measure distances, time intervals and velocities, and the sham-theory will now predict imaginary velocities.

But not all is lost. The physicists of the planet simply amend the theory and set c in the Lorentz transformation equal to the velocity of sound (in air at 0°C and 1000 mbar pressure). The Second Postulate now checks out beautifully for every velocity up to that of sound; but one day a scientist discovers the photocell and the existence of light, and the amended theory is refuted by the velocity of light signals.

So they amend the sham-theory once more and set c in the Lorentz transformation equal to the velocity of light, and what do they get?

The Einstein theory in its full glory.

Perhaps you can now see what I am getting at. If we define space and time to cater to a constant velocity of water, the theory is refuted by sound signals; if we define space and time to cater to a constant velocity of sound, the theory is refuted by light signals; and if we define space and time to cater to the constant velocity of light. . . but there is nothing faster than light.

This implies that the Second Postulate may well be something that is not inherently true, but that is merely protected from refutation by the lack of a "messenger" velocity faster than that of light. This possibility — and with it the possibility that the Einstein Theory is merely an equivalence — gains weight when it is realized that the Second Postulate (from which the Lorentz transformation immediately follows) has never been demonstrated by *direct* experiment.

There is another point of interest associated with the logical flaw, alleged or genuine, of tampering with the fundamental concepts and in effect defining them by higher-order concepts — not to mention points of built-in irrefutability. As I will point out below, there is, 80 years after the Einstein theory made its appearance, a sizable community of scientists who have not accepted it. And there is a far larger group of scientists who feel a pronounced distaste for it, though they shrug it off and accept the theory because there is no viable alternative. (Most scientists, of course, are in a third group: they never get deeply into the Einstein theory and "accept" it as I accept the theory of the genetic code and other theories outside my expertise.) It is my belief that this distaste stems from the opposition, conscious or not, to tampering with fundamental concepts such as time and simultaneity.

But no, we are told, the reason why people have difficulty with the Second Postulate, and hence with the rest of the Einstein theory, is quite simple. What prevents a few cranks, mavericks and flat-earthers from accepting such an unorthodox view of space and time is their inability to accept anything that is different from the world they are used to.

Then why are there no cranks rejecting the existence of atoms that nobody has ever seen? Why are there no "underground" scientific journals doubting the validity of thermodynamics? (There are several doubting the Einstein theory.) Why does the quantization of energy raise no hackles in a world in which all energy varies smoothly from a fly's sneeze to a 100,000 megaton-equivalent volcanic eruption?

It is, of course, the exact opposite that is true: not only physicists, but people in general *love* phenomena that are quite different from the world they are used to. They spend hard-earned money for a toy gyroscope just to see it balance on a piece of string when it really "ought to" fall off, and they are doubly fascinated when they see that it is no swindle. What they do not like is being asked to abandon reason: they grow wary when they sense a logical flaw.

They would be offended by a theory defining a straight line in terms of a rectangle, especially if its area is dependent on the state of the student contemplating it.

There is a counterexample to people's wariness of logical flaws: the Principle of Relativity itself, which never has any trouble being accepted. It is quite misleading to call the Einstein theory "the" theory of relativity, a name that I will not use. Einstein did not discover the principle, which was known to Galileo, though he did not explicitly state it. It was explicitly stated, though not under that name, by Newton in the *Principia*: Corollary V, Book 1, says *Corporum dato spatio inclusorum iidem sunt motus inter se, sive spatium illud quiescat, sive moveatur idem uniformiter in directum sine motu circulari* – "The momenta of the bodies included in a given space are the same, whether that space is at rest or whether it moves uniformly in a straight line without rotation."

There was no electromagnetics then; all of physics (then called "natural philosophy") consisted of mechanics and optics, the latter – in either the corpuscle or wave theory – considered to obey mechanical laws. Since all of mechanics can be reduced to momenta of bodies, Newton's statement surely is an explicit 17th century formulation of the relativity principle, which is today often stated as "the laws of physics hold equally well in all inertial frames."

Newton's belief in a system of absolute rest, based on considerations of *accelerated* (rotational) motion may have been unnecessary, but it did not contradict the principle of relativity valid for uniformly moving systems (inertial frames) which he had thus stated.

Let me take this opportunity to dispel another myth, namely that Einstein's theory contradicts Newton's Laws. The statement that force equals mass times acceleration was put in Newton's mouth posthumously: there is no place in the *Principia* where Newton makes such a statement. He always writes about the rate of change of momentum (*mutatio motus*, or "change of motion," the latter defined as the product *quantitatis materiae et velocitatis*). In present notation – the *Principia* make their case by geometry – Newton never took the m out of the parentheses in $d(mv)/dt$, for he was too careful a man to ignore the possibility that inertial mass might be variable. When Einstein introduced velocity-dependent mass explicitly, he did not have to change one iota in Newton's Laws of Motion for any part of his theory; that he developed it in contradiction to them is one of the numerous fables surrounding the Einstein theory. (Newton's law of gravitation is not, of course, one of the three Laws of Motion, nor does it have their generality and fundamental significance.)

But how relativistic is the Einstein theory, "the" theory of relativity?

If the laws of physics are conserved in all inertial frames, one would expect that it makes no difference whether an electric charge moves through a stationary magnetic field or a magnetic field sweeps past a stationary charge. The reason for this expectation, I submit, is our unperverted subconscious which says that magnetic fields and charges interact all by themselves, without the benefit of observers. But that is not what the Einstein theory says. A charge moving through a uniform magnetic field is acted on by a force; but a moving *uniform* magnetic field (which has no space or time derivatives, and therefore cannot induce an electric field) does not affect a stationary charge, for "moving" and "stationary" is defined with respect to the observer, not with respect to the field.

Even stranger, in the Einstein theory a moving charge does not act with the same force on a stationary charge as the stationary charge acts on the moving one. (The observer "sees" the moving charge with its electric field intensifiied in the direction perpendicular to the velocity, but the field of the stationary charge is unmodified.) Action and reaction are therefore no longer equal and opposite when the charges are interacting at a distance and not actually colliding at one point in "space-time." Only erudite Einsteinians are aware of this, and their answer is "So what?"

So what we have, if we believe in an objective reality unchanged by observers' perceptions, is a theory that fulfills the principle of relativity by distorting space and time in order to enforce the validity of laws expressed in terms of observer-referred velocities.

Why, then, have scientists universally accepted the Einstein theory?

They haven't. Most scientists have not studied it beyond a freshman course. Among those who have, most do accept it without reservations. But some turn away in queasiness — and in silence, for they have nothing better to turn to.

It is true that among the heretics there is a sizable percentage of cranks and simpletons; but there are others. There are those who lack mathematical training and simply feel that the Second Postulate does not square with common sense. And there are also, to this day, some rebels of academic standing whose grumblings can occasionally be heard in public.

Louis Essen, director (now retired) of the Time and Frequency Division of Britain's prestigious National Physical Laboratory, and a physicist of international renown, writes "A common reaction of experimental physicists to the theory is that although they do not understand it themselves, it is so widely accepted that it must be correct. I must confess that until recent years this was my own attitude." His analysis [Essen 1971] finds the theory self-contradictory.

Prof. Thomas G. Barnes, Professor Emeritus of Physics at the University of Texas, writes "It is time to return physics to a philosophy that puts physical reasoning ahead of blind faith in relativistic concepts that lead to nonsensical contradictions." [Barnes 1983.]

The late Herbert Dingle, Professor of the History and Philosophy of Science at University College, London, was originally an enthusiastic supporter of the Einstein theory, but in his study of the theory he found flaws and turned against it in numerous articles and a book [Dingle 1972].

Burniston Brown, retired Reader (Associate Professor) in Physics at University College, London, is the author of a recent book [Brown 1982], which makes the case for retarded action at a distance as an alternative to the Einstein theory. (So does the present book, but giving more emphasis to the effects of the aberrational component of the retarded force.)

These are but four of a sizable list of contemporary or recently deceased Einstein critics, and no offense is intended to those not listed here.

But perhaps no less impressive are the names of some of Einstein's opponents in his own time, and I do not mean the "natural" enemies of any new theory — the mediocre fossils who are threatened with having to unlearn a lifetime's investment in the old theories. I mean the names of those whose work is closely associated with the theoretical basis or experimental verification of the Einstein theory, but who — and this may come as a surprise to many — vigorously opposed it.

Hendrik Antoon Lorentz, author of the Lorentz Transformation, would have nothing to do with the Einstein theory and opposed it until his death in 1928.

Herbert E. Ives of the Ives-Stilwell experiment not only seethed in his personal correspondence over Einstein's contradictions, "guesses" and "hunches," [Hazelett and Turner, 1979], but also had the stature to be given space for his heretic attacks on the Einstein theory in established scientific journals as late as 1953, the year of Ives' death. And the incomparable Albert A. Michelson of the Michelson-Morely experiment remained doggedly faithful, until his death in 1931, to the "entrained ether" theory (with which, indeed, that experiment was perfectly compatible).[1]

Why, then, can objections to the Einstein theory be published only in the "underground" scientific press?

Because they merely show that there may be something radically wrong with the theory; but they have no full substitute to offer.

[1] Referring to the experiment, Michelson is said to have quipped "I created a monster." I have found no confirmation of this in the two biographies of A.A. Michelson that I have read, one of them by his daughter [Livingston 1973]. On the other hand, both books are somewhat apologetic about his refusal to accept the Einstein theory, and it could be that they did not want to throw even more "bad" light on him. But if the remark is apocryphal, it is well invented.

In a world where every possible experimental verification has shown uncanny agreement with the results predicted by the Einstein theory, such approaches will only get you a smile and a shoulder shrug. To beat the Einstein theory, it is not good enough to provide an alternative that does equally well; you have to show that it can do better.

Can it be done?

Part One

Einstein Minus Zero

1.1 The Static Inverse Square Law

Consider Newton's Law of Gravitation. For low relative velocities of the two interacting bodies ("low" velocity meaning here, and in the rest of the book, negligible compared with the velocity of light) it is quite uncontroversial; in polar coordinates, with the origin at the center of mass of one of the two bodies, and $\mathbf{r}_0$ the unit radial vector, it says that the force between the two masses is

$$\mathbf{F}_g = -\frac{\Gamma m_1 m_2}{r^2}\mathbf{r}_0 \tag{1}$$

where $\Gamma = 6.67 \times 10^{-11}$ Nm^2/kg^2 is the gravitational constant, and the rest of the formula, in fact the rest of this book, is also in SI units; the minus sign says that the force is directed *against* the unit vector $\mathbf{r}_0$, i.e. attractive.

The inverse square of a distance from a point is indicative of something – a force – emanating from a source at that point. We will assume that it propagates with the velocity of light c. We know from experience that in the electric analogy of (1) this is the case: for example, if we remove (discharge) the charge, the removal of the force at a distance r is delayed by a time r/c. We assume (with Einstein and practically every other gravity theoretician) that the same holds for gravity: that if we were able to "dismass" a mass as we are able to discharge a charge, then the result of this (or any other) modification would reach the field at a distance r only after a delay of r/c, the disturbance of the field traveling outwards with a velocity c.

This is quite a conventional assumption. It not only emerges from the Einstein theory, but it was also made by the late 19th century classics; in fact, it was made even earlier by Pierre Simon Laplace himself in Book 10, Chapter 8, of his *Mécanique céleste* (publ. 1799-1825). With no electromagnetism to go on, Laplace could not have foreseen that the velocity of propagation was that of light, but he explicitly worked with a velocity of propagation of the gravitational force.

The velocity c with which the force propagates from its source is measured with respect to the source, and this again is uncontroversial, for there are only two static bodies, and the interaction is that of one body in the field of the other. For this static case all theories — propagation, emission, ether, and the Einstein theory (with the observer located on one of the two masses) — yield the same result, and there is nothing substantial to determine the deeper nature of the mechanism that transmits the force with velocity c.

This deeper nature will not be needed in the following; nevertheless, it is intriguing to contemplate the *product* of the masses (charges) in the Newton-Coulomb Law. This implies that gravitational or electric attraction is a force quite unlike, say, the force stretching the rope in a tug-of-war, where the tension is proportional to the *sum* of athletes on each side. Masses and charges are evidently not team players: they interact individually, each particle of one body with each particle of the other.

A possible interaction that exhibits such uncollectivist behavior is a wave emitted by one source and interacting with all similar sources struck by its wave fronts. If such waves are the solution of the wave equation for the force field, they need no ether or particle flow or medium in which to propagate (though they do not contradict any of them); they are simply a wave motion of force in unspecified form.

Analogously, the Coulomb Law for the force between two electric charges is

$$\mathbf{F}_e = \frac{q_1 q_2}{4\pi\epsilon_0 r^2}\mathbf{r}_0 \tag{2}$$

where $\epsilon_0 = 8.854 \times 10^{-12}$ F/m is the free-space permittivity; if the charges are elementary (electronic, $q = 1.6 \times 10^{-19}$ C) but have opposite signs, then the constant $K = 2.32 \times 10^{-28}$ Nm^2.

This force again propagates outward from its two sources with velocity c, and this time we may regard the hypothesis as experimentally verified, at least to the extent that the force ceases to act with a corresponding delay when its source is removed.

1.2. The Velocity of Light: With Respect to What?

When light is emitted by a source moving uniformly through a vacuum, its velocity is constant; but with respect to what?

With respect to all observers, regardless of their velocities relative to the source, says the Second Postulate of the Einstein Theory. This is today the generally accepted answer despite the absence of a *direct* proof and despite the objections pointed out in the Introduction.

Before the advent of the Einstein theory, it was generally believed that light propagated in an all-pervading "luminoferous" medium, the ether. The velocity of light was constant with respect to the ether, just as the velocity of sound is constant with respect to the air in which it propagates, even though the source and the observer might be moving with different velocities with respect to the air.

There were, however, two varieties of the ether theory. In the first, the earth and other objects moved through the ether without affecting it, so that the velocity of light with respect to an observer moving through the ether was $\mathbf{c} - \mathbf{v}$, where $\mathbf{v}$ was the velocity of the observer, and both velocities were measured with respect to the ether.

In the "entrained" ether theory, the earth dragged the ether in its neighborhood along as it moved round the sun. The velocity of light would therefore be constant in all terrestrial laboratory experiments (including those made with starlight), since the ether was at rest with respect to the laboratory.

In the "ballistic" theory of light, whose main exponent was the brilliant young Swiss physicist Walther Ritz (1878-1909, died at age 31), it was assumed that the velocity of light is constant with respect to its source, like bullets from a machine gun on a moving train. It did not need an ether.

The alternative to the Second Postulate that I will work with is that the velocity of light is *constant with respect to the local gravitational field through which it propagates.*

The reason for this assumption is the reason for *all* assumptions in physics: it is supported by all the available experimental evidence and contradicted by none — as I hope to show in the following sections.

Let me first explain what is meant by "with respect to the local gravitational field."

As in any other conservative vector field, any point of a gravitational force field is defined by the line of force and the equipotential passing through it; its coordinates can therefore serve as a standard of rest. This approach will yield the correct result, though it throws no light on the physical mechanism involving it.

Alternatively, we may *think* of light as a disturbance of the gravitational field itself (something like sound, which is a disturbance of a pressure field); this will again yield the correct result, but there is no evidence whether this is a physical reality.

The "local" refers to light propagating through gravitational fields moving with respect to each other, as is the case for the planets, the sun and the stars. If the sun is the rest-frame, light from a terrestrial source would first move with a velocity $\mathbf{c}+\mathbf{v}$ (where $\mathbf{v}$ is the orbital velocity of the earth, about 30 km/sec) in the dominant terrestrial gravitational field, and then with velocity $\mathbf{c}$ in the rest frame. In the transitional region there would be a transitional velocity, marked by the properties of most transients: difficult and of secondary importance.

Beyond this simple consequence of Galileian relativity, the experimental evidence (bending of light rays in a gravitational field) suggests that the velocity of light varies with the intensity of a gravitational field; this is not surprising, since *all* cases of wave motion show a velocity dependence on the properties of their environment (the index of refraction). It is, however, a minor point to which we will not return until Sec. 1.11.

There is also hard experimental evidence that the velocity of light remains constant with respect to the earth's gravitational field, but not with respect to the earth rotating in it; this will be discussed in Sec. 1.3.7.

The assumption that the velocity of light is constant with respect to the local gravitational field is one that may raise many hackles as a *conceptual* formulation, but as an experimental fact it is not at all absurd:

First, it satisfies the relativity principle without attempting to redefine space and time. Like waves on the water of a stream flowing into a river and into the sea, light travels with different relative velocities through a vacuum in the terrestrial field, through that in the solar field, and through that of the fields that lie beyond; none of them is privileged or at absolute rest. If inertial frames are related to each other by the Galileian transformation, and time flows at the same rate in all of them, the laws of physics will hold equally well in all of them, as will be shown for optics and electromagnetics in the following, and as is surely obvious for the velocity of light by itself, without regard to its electromagnetic nature.

Second, there is a rarely noted, but nevertheless firm, precedent of an electromagnetic quantity that depends on a velocity with respect to uncharged matter (the source of gravitation). It is the magnetic field, not as it appears in thought experiments by this or that theory, but as it is measured in the macroscopic world. It is too weak to be measured unless the electric field of the moving charges is first neutralized, as is the case when a current flows in an overall neutral conductor. This is no new assumption, but a consequence of perfectly orthodox (including Einsteinian) electromagnetics, as will be pointed out in more detail in Sec. 1.4.

Third, this assumption cannot *experimentally* contradict the Einstein theory, for no observer or measuring instrument has ever traveled through a gravitational field with a velocity comparable to that of light — certainly not in uniform, rectilinear motion. In the cases where the motion was rotational, i.e. in Sagnac-type experiments (rotation of a double interference loop), the evidence supports both the present assumption and Einstein's general theory for rotating systems.

Thus, in all optical experiments supporting the Einstein theory, the observer was always nailed to the gravitational field of the earth; on the other hand, the

Michelson-Gale experiment utilizing the earth's rotational velocity (Sec. 1.3.7), which did register a fringe shift, is explained by the Einstein theory as a Sagnac-type experiment (an argument that can also be used for satellites). In these cases using the earth's rotational velocity, both assumptions lead to the observed result; however, Einstein's general theory, valid for accelerated frames, is mathematically so complicated and physically so opaque that only a comparatively small circle of specialists has mastered it.

Let us then quickly run through the crucial experiments of a purely optical character, that is, those that make no use of the electromagnetic nature of light, but treat it simply as something that moves with a measurable velocity and that is capable of interfering with itself. These are the experiments that do not in any way rely on electromagnetic inferences — such as those based on the tacit assumption that the expression for the Lorentz force remains valid at high, observer-referred velocities. This group of purely optical experiments thus excludes those involving charged particles.

The second, electromagnetic type of experiment can also be characterized by another property: it always involves the *square* of the quantity $\beta = v/c$, whereas the purely optical experiments are most often limited to first-order observations, i.e. to observing quantities depending linearly on β.

This makes the purely optical evidence not only more easily obtainable, but also less dependent on possibly flawed conclusions, and we shall examine it first.

1.3. The "Purely Optical" Evidence

Among the experiments that treat light simply as something propagating with a measurable velocity without reference to its electromagnetic character, we will examine the crucial ones performed with moving sources (including moving mirrors and moving media of transmission). By "crucial" I mean those helping to support or reject one of the four competing theories – ether, ballistic, gravitational, or Einstein's Second Postulate.

1.3.1. Aberration

In 1728, James Bradley (1692-1762), then Savilian Professor of Astronomy at Oxford, sent the Astronomer Royal (Newton's good friend Halley) an *Account of a new discovered motion of the Fix'd Stars*, noting that a star in the constellation of the Dragon crossed the meridian more to the south in the winter of 1725-26 than in the preceding and following summers, an effect that could not be explained by parallax.[1] The effect, called aberration, is reminiscent of vertical rain leaving slanted tracks on the side window of a traveling car: while the star light travels through the telescope with velocity c, the telescope moves forward with the earth's orbital velocity v (about 30 km/sec), so that the ray passing through the telescope makes an angle of aberration ϵ with the true direction of the star.

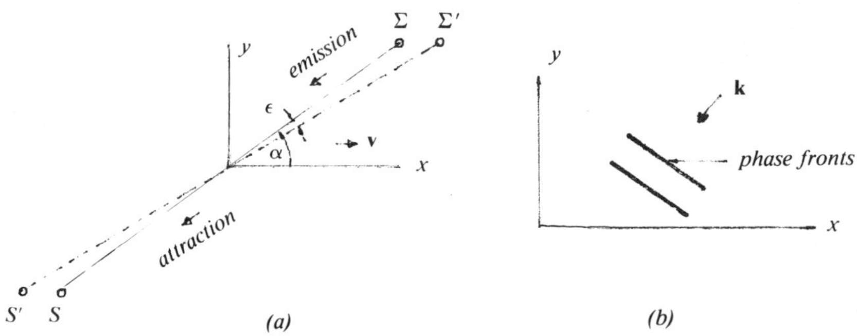

Aberration: (a) general geometry, (b) wave theory (see also p. 201)

Bradley's discovery was erroneously interpreted as a victory of the ballistic theory of light over the wave theory, probably for two reasons: the explanation by the ballistic theory (corresponding to the rain on the moving window) is much simpler; and Newton's criticism of the wave theory was misinterpreted as approval

[1] *Phil. Trans.* vol. 35, p.637 (1728). Whittaker [1910/62, p.94] notes that Roemer (the first to measure the velocity of light, using Jupiter's moons), in a letter to Huygens dated 30 December 1677, suspected the apparent displacement of a star and gave the correct explanation, thus preceding Bradley by half a century.

of the corpuscular theory. In fact, as those who have read the *Opticks* know, Newton refrained from endorsing either.

In reality, Bradley's discovery was not an *experimentum crucis*, for it can be explained satisfactorily by any one of the four theories. However, aberration plays a significant role in the theory to be proposed, so we will review it for later reference.

If **c** and **v** are, respectively, the velocities of light and of the object on which it is incident, both referred to the frame in which the source of light (or force!) is at rest, then by the ballistic theory, which treats light as it would machine gun bullets, we find the aberration angle ϵ by resolving the velocity of light in the telescope system into the direction of **v** in the star (Σ) system and into the direction perpendicular to it:

$$\tan(\alpha + \epsilon) = \frac{c \sin \alpha}{v + c \cos \alpha} = \frac{\sin \alpha}{\beta + \cos \alpha} \tag{1}$$

where $\beta = v/c$. After elementary manipulations this yields

$$\tan \epsilon = -\frac{\beta \sin \alpha}{1 + \beta \cos \alpha} \tag{2}$$

or neglecting second-order terms in β and ϵ we have approximately

$$\epsilon \approx -\beta \sin \alpha \tag{3}$$

The negative sign means that the aberration subtracts from the angle α and therefore deviates toward the direction of the velocity. However, this is true only of light or other agents that are *emitted* from a source. We shall soon have occasion to consider the aberration of an attractive force, such as Coulomb's or that of gravitation, that is directed *toward* the source (which might more accurately be called a sink). In that case the aberration is positive, so that it deviates *away* from the direction of the velocity. This is immediately apparent by noting that the *attraction* by the sun *S* in the figure is, as far as the geometry of aberration is concerned, equivalent to the *emission* of light by the fictitious star Σ.

The wave theory can, of course, do equally well, for the phase fronts or planes of constant phase in a system in which the star is at rest are given by

$$\Phi = \omega t - \mathbf{k} \cdot \mathbf{r} \tag{4}$$

where **r** is the position vector based at an origin fixed somewhere on the earth's orbit, ω is the angular frequency of the light, and **k** is the propagation constant with scalar value

$$k = \frac{\omega}{c} = \frac{2\pi}{\lambda} \tag{5}$$

If **k** is oriented as in the figure, we have

$$\Phi = \omega t + kx \cos \alpha + ky \sin \alpha \tag{6}$$

To find the direction of the phase fronts, we set $\Phi - \omega t = \text{const}$ (that is, we look for the locus of a constant phase Φ at a fixed time t), yielding the family of planes

$$y = -x \cot \alpha + \text{const} \tag{7}$$

which is not surprising, since the phase fronts are perpendicular to the direction of propagation $y = x \tan \alpha$.

But substituting $t = x/v$ in (6), we find the family of phase fronts as

$$\frac{v\Phi'}{\omega} = x(1 + \beta \cos \alpha) + \beta y \sin \alpha = \text{const} \tag{8}$$

or

$$y = -\frac{x + \beta \cos \alpha}{\beta \sin \alpha} + \text{const} \tag{9}$$

Comparing (7) and (9), we see the factor multiplying x in (9) plays the role of the cotangent of the same angle, namely the aberration angle by which the planes of equal phases shift when $t = x/v$ instead of $t = \text{const}$. Hence

$$\cot \epsilon = -\frac{1 + \beta \cos \alpha}{\beta \sin \alpha} \tag{10}$$

which is identical with (2).

The same derivation is valid for the gravitational field theory. There is an aberration as the ray from a star enters the gravitational field of the sun if it has a velocity with respect to the star; and there is a further aberration as the ray passes from the gravitational field of the sun to that of the earth – for simplicity I am replacing a continuous transition from the dominance of one field to the other by a sharp discontinuity, assuming that a more careful treatment would introducc time-consuming details, but no substantial modifications. We are then back to the preceding derivation: we simply take the entire blob representing the earth's gravitational field instead of considering it a point as above. This will be discussed in more detail in connection with Airy's experiment below.

The aberration formula derived by the Einstein theory agrees with (1) to first order in β. (For the orbital motion of the earth, $\beta = 10^{-4}$; second-order verification would thus require the measurement of an angle with a precision of 1 in 100 million – supporting the assertion in the introduction that the crucial verifications of the Einstein theory always rely on electromagnetics, the only field where significant values of β^2 are achievable.) It is derived directly from the Lorentz transformation and is a property of the different values observed in the space-times of the two inertial frames.

However, here and in the following I will not reproduce the Einsteinian derivations. There is an abundance of available books making the case for the Einstein theory, and the observant reader may have detected that this is not one of them.

1.3.2. Fresnel's Coefficient of Drag

Augustin Jean Fresnel (1788-1827) was a rare genius, whose work should fill one with humility not only because of its volume and significance, but because of the pause-giving thought that virtually all the results derived by him remain valid to this day even though they were derived from the concept of an elastic, compressible ether.

This circumstance is striking enough in such cases as the reflection and transmission coefficients, diffraction formulas, and path clearance criteria without which contemporary microwave relay lines could not be designed. But it is almost uncanny in the case of the dragging coefficient for moving media: Fresnel derived it, without experimental evidence, from the idea of a compressible ether being partially dragged along by a body moving through it. Yet the formula he thus derived in the "wrong" way has not only stood the test of time, but it also had a significant effect on the acceptance of the Einstein theory: the only competing theory at the time, Ritz's ballistic theory, was unable to provide an explanation for the experimentally confirmed dragging coefficient, whereas Einstein obtained it as the leading term of the series resulting from his velocity-addition theorem. (The double-star argument against Ritz, apart from being flawed, had not yet appeared. On the other hand, Ritz apparently did not know about the work of Hoek, half a century earlier, whose "etherless" derivation of the drag coefficient he might have used as a defense.)

The Fresnel coefficient of drag describes the velocity of light in a moving material medium.

The refractive index n of a medium is defined by

$$n = \frac{c}{c_m} \tag{1}$$

where c_m is the velocity of light in that medium when it is at rest. If now a transparent medium moves with velocity v with respect to a fixed system of coordinates, then the velocity of light in that medium, which is c/n at rest, is increased by an amount δv when measured with respect to the same system:

$$c'_m = \frac{c}{n} + \delta v \tag{2}$$

where

$$\delta = 1 - \frac{1}{n^2} \tag{3}$$

is Fresnel's coefficient of drag.

This expression will be derived in Sec. 1.10.2 from purely electromagnetic considerations using the Galileian transformation; but in this non-electromagnetic survey it is important to note that neither ether nor Einstein are needed to obtain it; it was derived by Dutch astronomer M. Hoek on using the result of a little known experiment that he performed more than a century ago [Hoek 1868][1]

Hoek compared the velocity of light in glass and vacuum (air), orienting a glass rod east-west, so that one of the two counter-running interference loops (see figure) passed through the glass in the direction of the rotating earth (west-east), and the other against it. He found no fringe shift when he reversed the direction of the entire apparatus.

I repeated the experiment with laser light in 1970 at the Engineering Center of the University of Colorado in Boulder, about 800 m north of the 40th parallel, where the rotational velocity of the earth is $v \approx 355$ m/sec, so that $\beta = 1.18 \times 10^{-6}$. Since laser light allows an optical path difference of thousands of wavelengths, this set-up allows the paths in air and glass to be compared directly without a double

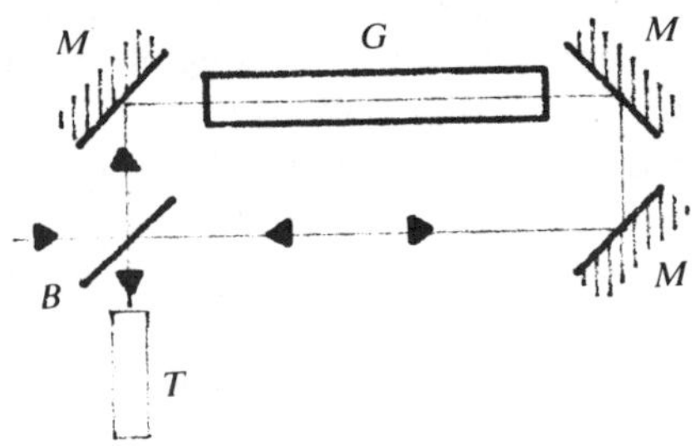

Hoek's experiment [1868]

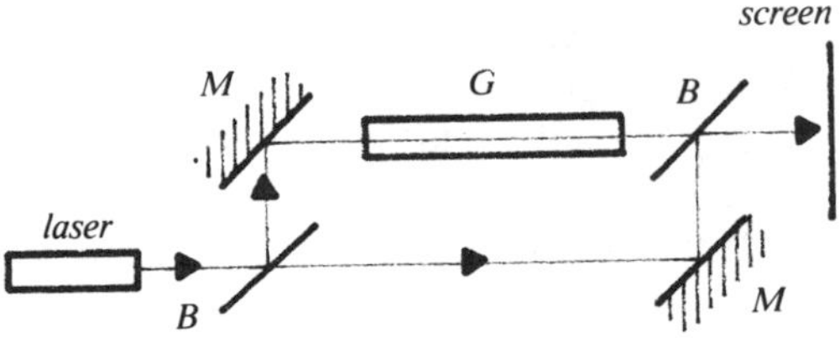

Repetition by Beckmann in 1970

loop; moreover, the whole arrangement was mounted on a styrofoam float in a tank of water, so that the fringes could be observed as the interferometer was rotated through 180°. The fringe shift after reversal was the same as Hoek's: none.

The experiments themselves are of no great importance; Hoek's is not widely known, and I never published mine.

What is of fundamental importance, however, is the way in which Hoek [1868] derived the Fresnel drag coefficient without recourse to an ether. Most textbooks never mention Hoek, and imply that there are only two ways to derive the coefficient of drag: Fresnel's using an elastic ether, and Einstein's using the first term in the power expansion of the velocity addition theorem.

[1] A paper published by an obscure Dutch journal in 1868 is not the easiest to obtain, and I am greatly indebted to my friend and countryman Ir. Pavel Dolan, now of Rijswijk, Netherlands, for sending me a photostat.

The essence of Hoek's derivation (applied to my simpler experiment) is the following. If l is the length of the glass rod (the other parts of the path cancel without affecting the result) with refractive index n, and the rotational velocity of the earth is v, then the time difference of the two phase fronts, with the light going east to west against the rotation of the earth, is by Galilean relativity

$$\Delta T = \frac{l}{c/n - \delta v - v} - \frac{l}{c+v} \tag{4}$$

and when the apparatus is reversed for the light to travel east to west, the time difference is given by the same expression with the sign of v reversed. Since there is no fringe shift, the two ΔTs must be equal, and on neglecting second order terms in β, we obtain (3).

Hoek's double loop is subject to the same procedure, and again yields the Fresnel coefficient with nothing but classical relativity — without ether or Einstein.

1.3.3 Fizeau's and Airy's Experiments

Fresnel had derived (2) and (3) of the preceding section without experimental evidence, and his hour of triumph came in 1851, 24 years after his death in 1817, when Fizeau confirmed the formula by an interference loop immersed in pipes running water with and against the loop branches at a velocity of $v = 7$ m/sec. The experiment is too well known to be described in detail, especially since the discussion of Hoek's derivation above shows that Fizeau's experiment is no more crucial to support of the Einstein theory than support of the ether. (Fizeau's experiment preceded Hoek's, but because of Hoek's derivation, I listed it first.) However, we can well imagine how Fresnel's confirmed prediction caused the scientific community to be unshakably convinced of the physical reality of an elastic, partially entrained ether that only cranky mavericks could doubt — just as today it is convinced of the physical reality of Einstein's theory for exactly the same reason: have its predictions not been vindicated by experiment?

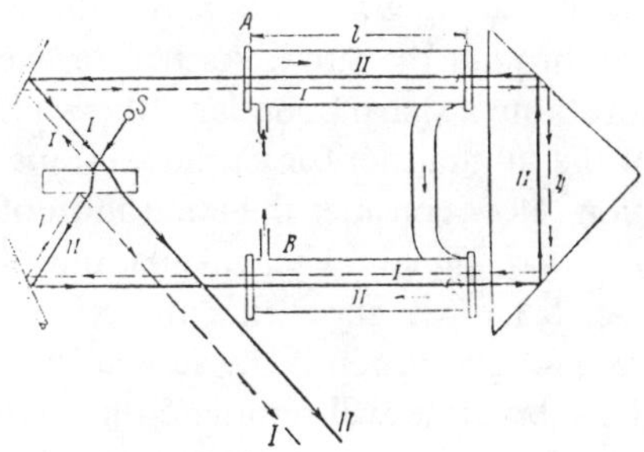

Fizeau's experiment (1851)
S source of light, *I* interference loop running with the water, *II* interference loop running against it. The length l was 1.5 m, the velocity of the water 7 m/sec.

Yet Fresnel's ultimate triumph did not come until 1871, when Sir George Airy (1801-92) performed an experiment that Fresnel, by then 44 years dead, had proposed: repeat Bradley's experiment using a telescope filled with water. By the ballistic theory, this would result in replacing c by c/n in (1), Sec. 1.3.1, and β by β/n; when the effect of refraction is considered together with that of aberration, the aberration angle is found proportional to n^2, which (for water) would increase the aberration angle by 7.6°.

Fresnel, however, concluded that the aberration angle would not change: the velocity of light would slow in the water, but the coefficient of drag in the water moving through a stationary ether would increase the velocity, and in calculating the aberration angle, the two effects would exactly cancel, so that the angle of aberration would be independent of the index of refraction. He wrote to Arago in 1818, "Although this experiment has never been performed, I do not doubt that it will confirm my conclusion."[1]

Airy's experiment found no change in aberration and once more fully confirmed Fresnel's concept of an elastic, partially entrained, compressible ether.

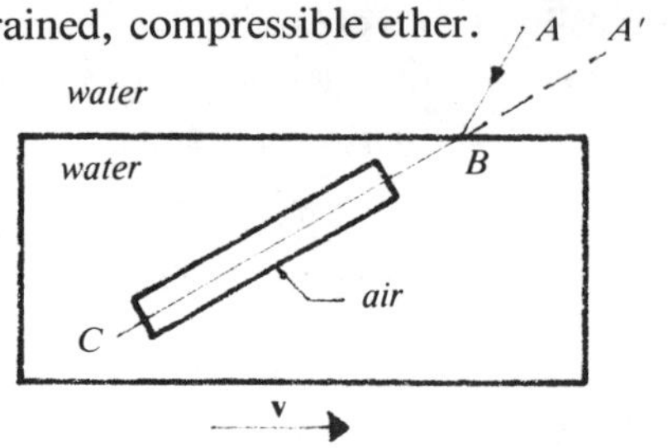

Analogy of Airy's experiment: in a water-filled submarine moving through the water, the same aberration will set in when sound waves are observed at the end C of a tube, no matter whether the tube is filled with air or water.

To see that Airy's experiment does not contradict the gravitational-field assumption, consider the analogy of a water-filled submarine moving through stationary water, representing the earth's gravitational field moving through the sun's. As before, we substitute a sharp boundary (the walls of the submarine) for a gradual transition. Now consider the aberration of a *sound* signal. The calculation of Sec. 1.3.1 will show that there is indeed an aberration if the submarine is adopted as a rest frame, since the walls of the submarine will reradiate the sound wave in the direction of arrival from the outside water. (If there are no walls marking an abrupt transition, the change in direction comes about in a curve rather than a "corner" at B but makes no difference to the end effect.) If we now slow the sound with an air bubble representing the telescope (remember that unlike light, sound travels *faster* in the denser medium), what will happen to the direction of the sound ray in the bubble inside the submarine?

Nothing whatsoever: the aberration has already changed the direction at the interface of moving and stationary medium, and there can be no additional aberrational change inside the submarine. Indeed, if we could detect a change in direction, we would be able to detect uniform motion without looking out of the window or otherwise referring to a rest standard: we would resurrect absolute rest and kill the relativity principle.

[1] Fresnel's derivation is now only of historic interest, and I do not want to waste space on it. Readers will find it well summarized in [Whittaker 1910], and those who read Russian will find a much simpler description of Fresnel's argument in G.S. Landsberg's *Optika* (Moscow 1952), pp. 363-4, showing how Fresnel worked with a compressible ether.

But would that not also protect the entrained ether theory from Airy's experiment?

Yes, it would – if one were to introduce the entrained ether as a confusing and unnecessary synonym for "gravitational field."

1.3.4 Double Stars and Other Objections to the Ballistic Theory

Consider the binary stars A and B revolving about a common center of mass and emitting light, as assumed by the ballistic theory, traveling at velocity c with respect to its source. Then in a system at rest with respect to (say) the common center of mass, the light emitted in the direction marked **c** by the receding star B is slower than that of the advancing star A, but the latter has a handicap h of up to the major axis of its orbit. Its faster light should, therefore, eventually catch up with its slower brother. This, it was claimed, would lead to several effects when these two stars are observed on the earth: the orbits viewed in such light should deviate from Kepler's laws [De Sitter, 1913], and the Doppler-shifted spectral lines of their light should double and triple. But no such effect has ever been observed, and this was used to reject the ballistic theory.

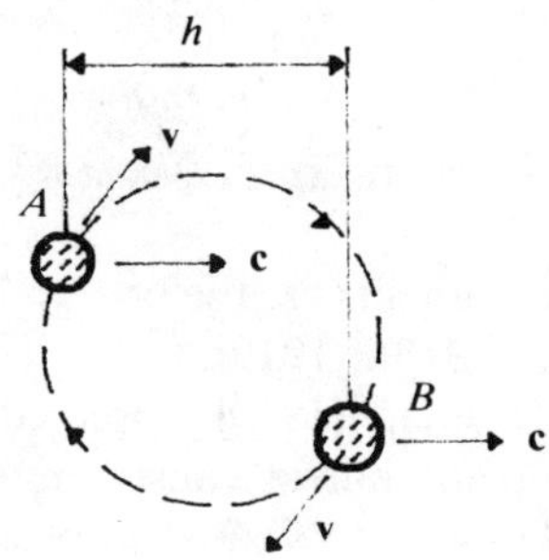

"Ballistic" light from star A in the direction **c** would eventually catch up with the slower light from star B.

The evidence from *laboratory* experiments does indeed refute the ballistic theory, as we shall see; however, the double-star argument is flawed for several reasons, of which only one is of interest here, because it illustrates the gravitational-field hypothesis. The alleged refutation tacitly assumes that the light emitted by double stars will remain constant with respect to the center of the double star from the moment of emission through the years or centuries of travel until it arrives at the terrestrial spectroscope. But from the point of view of the gravitational hypothesis this tacit assumption is false: the two light rays will indeed at first travel with different velocities, constant with respect to either star; but they will soon stabilize at a common velocity as the gravitational fields of the two merge into one; the velocity will change to a different, but again common, value as the light enters the next dominant field on its journey, for it is still c, but now with respect to a different field; and so forth until it enters the telescope of a terrestrial observatory, with no reason for any special effects.

There is, however, convincing laboratory evidence against the ballistic theory. There are several experiments disproving the theory if it is assumed that the velocity of light is not modified as it passes through lenses and is reflected by moving mirrors like a tennis ball bounces off a racket – with twice the velocity of the moving mirror added to the incident velocity. Among the more convincing

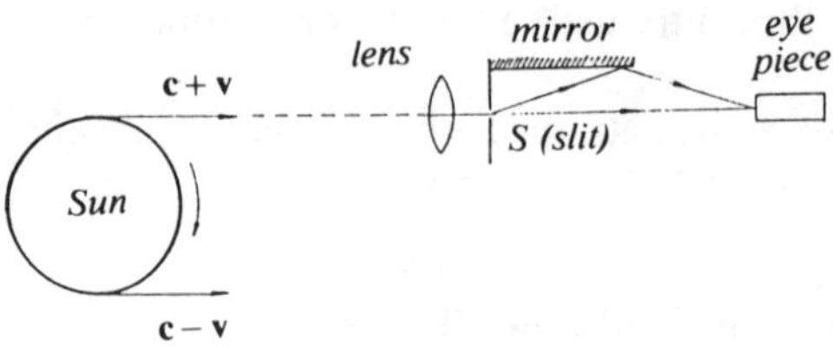

Tolman's experiment [1912]

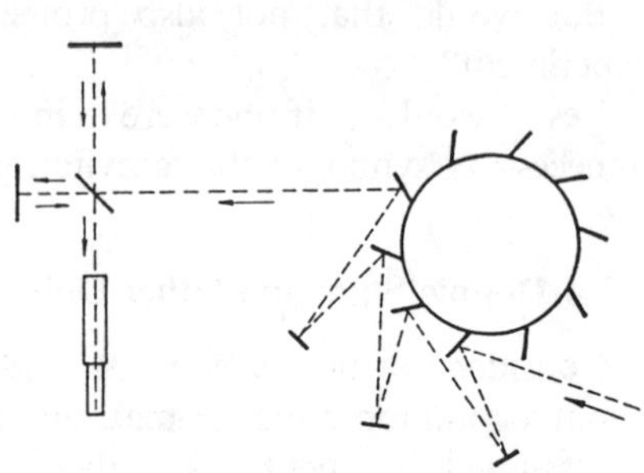

Majorana's experiment [1917, 1918a]

experiments are the one by Tolman [1912] and the two by Majorana [1917, 1918a and 1918b, 1919].

Tolman [1910] observed the limbs of the sun with a Lloyd interferometer (a plane mirror causing interference with the direct ray near grazing incidence). The sun, at its equator, rotates with a circumferential velocity of about 2 km/sec, so that by the ballistic theory the light emitted by the receding limb should have a lower velocity than that emitted by the advancing limb, and a fringe shift should therefore be observed on pointing the interferometer first at one limb, then at the other; but none was observed. The experiment was repeated with lavish equipment by Bonch-Bruyevich and Molchanov [1956], who were apparently unaware of Tolman's experiment 44 years earlier, for they make no reference to it.

Majorana [1917, 1918a] mounted 10 mirrors on a rotating wheel with a circumferential velocity of up to 70 m/sec and let the light, after traversing the indicated path, pass into a Michelson interferometer with unequal arms. The simple ballistic theory should have shown a Doppler shift in addition to the one predicted by either the ether or the Einstein theory, and the "tennis ball" version should have shown a further shift due to the stationary mirrors. But only the normal Doppler shift was observed. The same result was obtained when the mirrors were replaced with active sources (mercury lamps) [Majorana 1918b, 1919].

There is, however, one intriguing variant of the ballistic theory that is less easily refuted: the "reradiation" version, by which mirrors and lenses reradiate the received (ballistic) light with a velocity c with respect to themselves, so that if they are stationary, the light transmitted or reflected by them propagates with the same velocity as predicted by either the ether or the Einstein theory. Since most interferometers, such as Tolman's or Majorana's, have a lens at their entrance point, they become useless for refuting this theory. It fell to the Grand Master of experimental optics, Albert A. Michelson, to devise an experiment that tested and refuted all versions of the ballistic theory in one brilliant swoop (Sec. 1.3.6) — and I do *not* mean the well-known Michelson-Morley experiment, which does no contradict any of these versions.

1.3.5. The Michelson-Morley Experiment

There is no need to go through the well known Michelson-Morley experiment of 1881; it is available in any textbook, which probably also claims that it conclusively disproved the existence of an ether.

It did no such thing, of course; it was perfectly consistent with an ether entrained by the earth, and Michelson interpreted it that way, unabashedly holding to that view in his writings decades later, for example in his 1924 experiment to be discussed below. That year, Tomaschek [1924] repeated the Michelson-Morley experiment with starlight, lending even more support to both the Einstein and the entrained-ether theory.

Nor did it in any way refute the ballistic theory: it is obviously consistent with all of its variants. And it is perfectly consistent with the gravitational hypothesis. To regard this experiment as a proof of Einstein's Second Postulate is one of the ironies attached to it.

The importance of Michelson-Morley is above all historical, for its result stood in shocking contradiction to Airy's experiment of a mere 10 years earlier, which had put Fresnel's *partially* entrained ether theory (the scientific community thought) on solid rock. The puzzle was to what extent the ether, whose existence nobody doubted, was entrained; and when Einstein and Ritz more than 20 years later each appeared with a theory that denied the very existence of an ether, many must have thought "a plague on both your houses."

Apart from its historical importance from this point of view, the experiment was also a milestone in that the Michelson interferometer used in it was the first that could have detected a fringe shift of order β^2 rather than only of order β. As used in 1881, the shift was close to the limits of detectability, but the technique was later perfected.

It is thus ironic that this experiment, which is perfectly consistent with four out of the five theories discussed here, and which refuted nothing but the unentrained (or only partially entrained) versions of the ether theory, should be held up in textbooks as proof of the Einstein theory or disproof of classical physics.

But the saddest irony is that the name of Albert A. Michelson, in the minds of most Americans, should be linked only to this experiment, which was a minor gem in his stunning treasury. For Michelson was a superstar in the field of experimental optics; his wizardry has not been matched again by any single optician. His feats such as measuring the diameters of distant stars by interferometry have no place in this book, but even the two all but forgotten experiments discussed in the next two sections show the unmistakable hand of the *maestro*. It is depressing that Americans should know no more about this man who at age sixteen buttonholed President Grant near the White House to get into the Naval Academy ("I will make you proud of me if I get the appointment!") and who made good on his promise by bringing the United States its first Nobel Prize in 1907.

1.3.6 Moving Mirrors

To decide whether the wave or ballistic theory was correct, Michelson [1913] mounted two mirrors on the shaft of a motor as shown in the figure and made

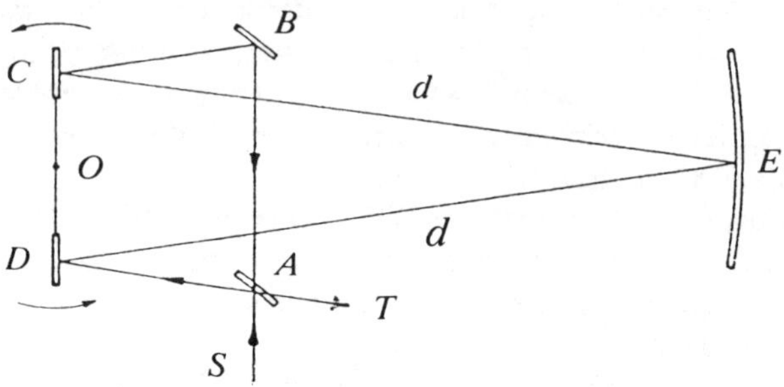

Michelson's experiment with rotating mirrors [1913]
C and *D* are mirrors mounted on a motor shaft rotating about *O*; the mirrors *B* and *E* are stationary. *S* source, *A* beamsplitter, *T* telescope.

them reflect the light in an interference loop. The length d from the stationary concave mirror at E to the motor shaft O was 6.08 m, the distance between the mirrors was 26.5 cm, and the motor speed could be varied continuously from 0 to 1800 rpm. Let $V=c-rv$ be the velocity of the light reflected from the mirror moving with velocity v, where r is an integer to be determined: if light propagates with constant velocity with respect to the laboratory, then $r=0$; if the mirror acts as a new source, $r=1$; if it is reflected like a tennis ball from a racket, $r=2$. An elementary calculation then shows that the fringe shift is

$$\Delta = \frac{v(T_1 - T_2)}{\lambda} = \frac{kdc}{v\lambda} \tag{1}$$

where λ is the wavelength of the light, the T_j are the time delays for the two directions round the loop, and k is 0, 4 or 8 for ballistic tennis-ball reflection, ballistic reradiation, and constant velocity with respect to the laboratory frame, respectively. The experiment gave $k=8$, showing the velocity of light, within the error of the experiment (about 2%) unaffected by the velocity of the mirrors.

"Assuming that the effect is actually nil," Michelson adds drily, "this interference method may be used to measure the velocity of light with an order of accuracy . . . of one part in 100,000."

Gentlemen, meet Albert A. Michelson: the length of his interference loop is more than 12 m, which is not trivial even with a laser; he kills three birds with one stone (or rather kills two and gives lasting life to the third); and he walks away with a new, state-of-the-art measurement method — all in one paper covering no more than 3½ small pages with large print.[1]

Michelson's little known experiment was the only one to refute the ballistic theory in its re-radiation version, but it was done in air. Therefore in 1964, Peter

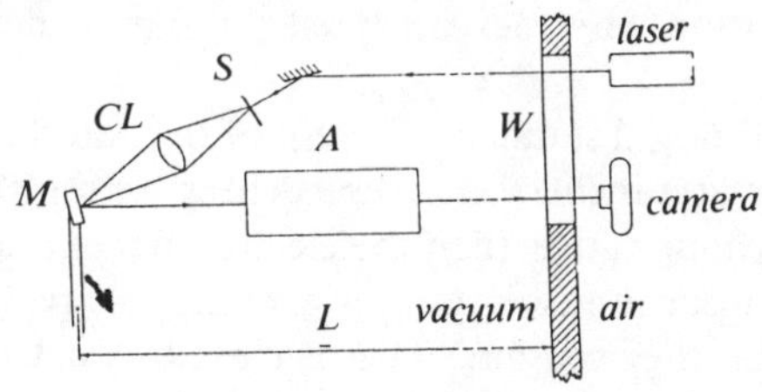

Test of the "reradiation ballistic" theory in vacuum [Beckmann and Mandics 1965]. *S* slit, *CL* collimator, *M* rotating mirror "accelerating" the light into Lloyd interferometer *A* and through the window *W*. Length *L* was 4.08 m.

Mandics and I performed an experiment in a chamber evacuated to 10^{-6} mm Hg [Beckmann and Mandics, 1964, 1965]. The basic idea was similar to Tolman's 1910 experiment (Sec. 1.3.4), but it was done with laser light, and the light was (by the ballistic hypothesis) "accelerated" by a mirror rotating in front of the slit, so that there were no lenses that could have "slowed" the light by reradiation. The ballistic theory predicted a shift of up to 0.7 of one fringe, but in fact we observed no shift as the speed of the mirror was increased.

With hindsight, this was fortunate. By the relativity principle applied to the ballistic theory, it makes no difference whether the mirror moves in air or whether a wind blows against the mirror. In the latter case we can use the coefficient of drag (the refractive index of air differs from 1 by an amount of order 10^{-3}), and if evacuating the air had changed the null effect, it would either have ruined the relativity principle or established that the refractive index of air is close to that of water.

With all this evidence against it (see also Sec. 1.3.4), I now consider the ballistic theory untenable. Other critics of the Einstein theory (see, for example, [Waldron 1977]) are still clinging to it, and I wish them luck, though I doubt they can undermine the Holy Grail from *that* direction.

---

[1] There is, in my opinion, only one man in the history of experimental optics to rival Michelson, and that is, once again, Sir Isaac Newton. Michelson worked without lasers, but Newton worked with his bare hands. The *Opticks* records his drawing of diffraction lobes clearly corresponding to those that Fresnel calculated three generations later with the aid of the integral named after him. Today, these lobes are easily demonstrated with a laser and a finely honed slit. What Newton used instead of a slit was the edge of a kitchen knife; and what he used for a collimated beam was the sun coming through a small hole in the barn which he had otherwise blacked out — in rainy England of all places.

There remains the puzzle of why, with the evidence available *at the time*, so many scientists either still adhered to a discredited ether theory, or accepted the Einstein theory rather than Ritz's non-ether ballistic theory, which explained not only the propagation of light from moving sources, but also the decrease of electromagnetic force on a moving charged particle (increase in inertial mass), inertia, and gravitation (including the advance of Mercury's perihelion). Equally puzzling is the question why they accepted the flawed double-star refutation so easily.

Herbert Dingle [1960], a staunch adherent of the ballistic (Ritz) theory, thought that the double-star argument was accepted because scientists "were prepared to sacrifice almost anything rather than the electromagnetic equations, and a reason for shaking off a nuisance rather than a genuine test between two equally valid possibilities was what they sought." That is close to what I will call the blunt side of Ockham's razor: when two theories become incompatible (such as Newton's and Maxwell's), we sacrifice the simpler one, for we have a heavier investment of learning (and a cherished chunk of snobbism) in the complicated one.

None of which is to say that Ritz was right; I believe the experimental evidence against the ballistic theory of light is now overwhelming (I do, of course, use the same idea as his *propagation des forces*). It is, nevertheless, shameful that this genius is forgotten by all but a handful of anti-Einsteinians: The Wiley/Interscience *Biographical Dictionary of Scientists* does not consider him worthy of inclusion among more than 1,000 scientists (such as Henry Ford), and a Soviet two-volume, 1000-page, fine-print encyclopedia of scientists (such as Trofim Lysenko) allots him 11 lines, mentioning only his work in spectroscopy and the "Ritz method" of solving variational problems.

1.3.7 The Michelson-Gale Experiment (1925)

The examples discussed above cover the main *types* of experiments on the velocity of light from moving sources or in moving media: there are more within the same type, but they do not essentially differ from the classes discussed above. All of them can be explained either by the Einstein theory, which assumes that the velocity of light is constant with respect to everybody everywhere moving at any velocity, and by the hypothesis that it is constant with respect to the gravitational field through which it propagates. The agreement between the two is not surprising, because the "everybody everywhere" (my expression, not Einstein's) has not been tested by anybody but observers at rest with respect to the earth's gravitational field.

There remains, however, a possible ambiguity: the earth's gravitational field moves with the earth as it travels along its orbit; but does it also turn with the earth about its axis, or does the earth revolve within its own field? In other words, with respect to what is the velocity of light constant on the *rotating* earth?

According to our hypothesis in Sec. 1.2, which regards the gravitational force as propagating from its source with a finite velocity, intuition suggests that the earth rotates within (with respect to) its own gravitational field, for once the force has been emitted, the earth rotates away under what it has emitted or what is propagated outward from it; the emitted or propagated agent is no longer under its control. Indeed, if the force remained under the control of its source after being emitted, we would need an additional hypothesis, including the requirement that the controlling message travel faster than the originally transmitted force; and while I will readily accept velocities higher than c for observers moving toward a source through the local field, I know of no evidence of a velocity higher than c with respect to the local gravitational field.

Fortunately, we do not have to rely on such intuitive reasoning, for it is confirmed by another Michelson symphony. As we have seen in the case of Hoek's experiment, the use of a moving material medium thwarts the intent of any experiment to measure a change of the velocity of light because the effect of motion is exactly canceled by the effect of refraction — that is the essence of the drag coefficient, and that is how Hoek derived it from his experiment. (This applies, of course, just as well to Airy's experiment, which uses aberration, i.e. orbital motion, rather than rotation about the earth's axis.) The effect of the earth's motion can therefore be detected only by comparing light *in vacuo* (or air) to light *in vacuo*.

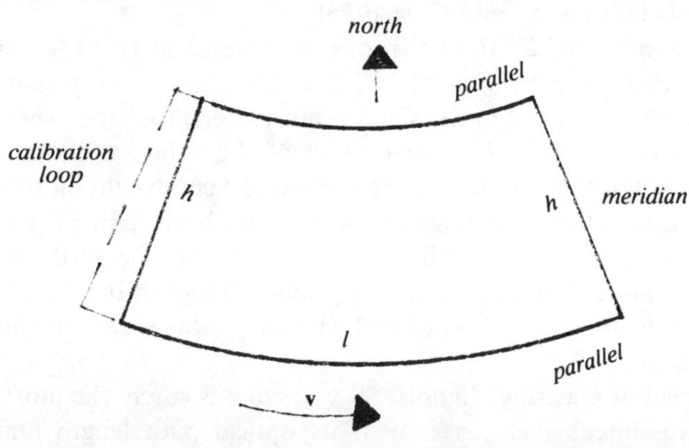

The Michelson-Gale experiment [1925] is based on the earth's slower rotational velocity at higher latitudes. An interference loop formed by a spherical rectangle will therefore show a fringe shift (compared with a rectangle with short east-to-west sides) because — in the classical conception — the velocity of light differs along the northern and southern sides of the rectangle.

This can be done by exploiting the variation of the circumferential velocity of the earth's surface, which rotates more slowly with increasing latitude: in a rectangular interference loop with east-west and north-south sides, the southern side of the rectangle will (in the northern hemisphere) rotate slightly faster than the northern side. Michelson proposed the use of such an interference loop as early as 1904: If v_1 and v_2 are the circumferential velocities of the earth at the northern and southern sides of the (spherical) rectangle with sides l_1, l_2, then the difference in time required for the two beams to complete the loop in opposite senses is

$$\Delta T = \frac{2l_2 v_2}{c^2 - v_2^2} - \frac{2l_1 v_1}{c^2 - v_1^2} \approx \frac{2(l_2 v_2 - l_1 v_1)}{c^2} \tag{1}$$

so that after some elementary spherical trigonometry the fringe shift is

$$\Delta = \frac{4lh\omega \sin \varphi}{c\lambda} \tag{2}$$

where ω is the angular velocity of the earth's rotation, l and h are the sides of the rectangle, φ its geographic latitude, and λ the wavelength of the light.

The slow rotation of the earth ($\omega = 73$ microradians/sec) necessitates a large area lh to give an appreciable fringe shift; and that in turn means taking the loop outdoors into the turbulent atmosphere. The experiment was not tried until 1923 near the Mount Wilson observatory, but the atmosphere and the resulting unsteadiness of the fringes were too much even for an artist like Michelson, who had hoped to save the expense of evacuated pipes. Using such pipes, the experiment was carried out successfully two years later, in the winter of 1924-25, at Clearing, Illinois [Michelson and Gale 1925].

The rectangle measured 2,010 feet from east to west and 1,113 feet from north to south, and was formed by straight and level 12-inch pipes connecting the four concrete boxes containing the mirrors and beamsplitters; the pipes were evacuated to half an inch of mercury. The wavelength of the light was 5,700 angströms (5.7×10^{-7} m). A total of 269 measurements was taken, usually in sets of 20 for given conditions (weather, exchange of mirrors and beamsplitters, etc.). On the "hypothesis of a fixed ether" (in Michelson's words), i.e. the earth rotating in a stationary ether without entraining it, (2) yields a fringe shift of 0.236; and the observed shift (yes, there was one) agreed with that value within the limits of the observational error.

The experiment at Clearing, Illinois, 60 years ago is surely the most grandiose interference experiment ever performed: its optical path length amounted to something like 10^{14} wavelengths, traversed with what was then considered "monochromatic" light: the light from a carbon arc passed through a filter.

Michelson's feat, which to my knowledge has never been repeated, is both a technical masterpiece, and one that provides fundamental insight into the optics and electromagnetics of moving sources. Yet it rarely makes it into the textbooks – certainly not into the introductory ones.

However, the result does not contradict the Einstein theory, at least not when the rotation rather than uniform translation of the coordinate system is invoked. The Einstein theory classes Michelson-Gale as a Sagnac-type experiment, so named after G. Sagnac, who in 1913 demonstrated a fringe shift by rotating an interferometer (with a polygonal interference loop traversed in opposite senses) at high speed; in such cases Einstein's general theory predicts a shift proportional to the angular velocity and to the area enclosed by the light path – not because the velocity of the two beams is different, but because they each have their own time. In the present case, the general theory also predicts the shift (2).

There is, nevertheless, one significant difference between the two explanations, and that is that (2) follows from the Galileian principle of in a few lines of high-school algebra, whereas Einstein's general theory does it with multidimensional complex tensors in space-time and non-Euclidean geodesics.

1.4. Magnetic Force and Gravitational Field

The force exerted by one charge by the field of another is given by the Lorentz force

$$\mathbf{F} = q(\mathbf{E} + \mathbf{v} \times \mathbf{B}) \tag{1}$$

where $\mathbf{E}$ is the electric field strength, and $\mathbf{v}_1$, $\mathbf{v}_2$ are the velocities of the two charges, the former the velocity of the charge q, the latter (to appear in a moment below) the velocity of the charge that produces the field through which the first is moving – and for the moment we leave it open with respect to what rest standard these velocities are defined. $\mathbf{B}$ is the magnetic flux density given by

$$\mathbf{B} = \frac{\mathbf{v}_2 \times \mathbf{E}_c}{c^2} \tag{2}$$

where $\mathbf{E}_c$ is the Coulomb field, i.e., the irrotational part of the electric field-strength $\mathbf{E}$. Relation (2) is a consequence of the Maxwell equations; it will be derived in Part Two, but for the time being we will just regard (2) as a definition of $\mathbf{B}$. The case of many charges moving in the field of many others then follows by superposition.

We define q as a scalar that modifies action at a distance.

Let us look at this in a little more detail. Two spheres of matter that do not have any charge will attract each other gravitationally by the inverse square law. When they are given an electric charge, the form of the inverse square law remains unchanged, but the scalar value (including sign) of the force between the two bodies will change. If the ratio of charge to mass is sufficiently large, the gravitational field may be neglected, at least locally. There is no action at a distance known to us other than either gravitational or electromagnetic, and both obey the same basic force law – certainly at rest, and presumably also in motion (the corresponding "gravimagnetic" field would usually be so weak as to escape direct detection). Defining charge or charge density as that which modifies the scalar value of a gravitational field, therefore, may be unusual, trivial or even inept; but it is perfectly consistent with what is usually understood by charge.

Now let us return to the force equation. Formally, equations (1) and (2) are the same as used in the Einstein theory, but there is a fundamental difference. The Einstein theory measures the velocities in (1) and (2) with respect to the observer, and the ether theory with respect to the ether. Before discussing with respect to what the two velocities are defined in the present theory, let me briefly go over some ground that is undisputed, yet not widely realized. In particular, few textbooks make the following point:

The magnetic force between moving charges is so small compared with the electric force between them that (today) *it is not measurable unless the latter is neutralized.*

Imagine an Einsteinian observer (at rest in the ether, to include that theory, too) observing two rigid rows of electrons, or negatively charged tennis balls, moving

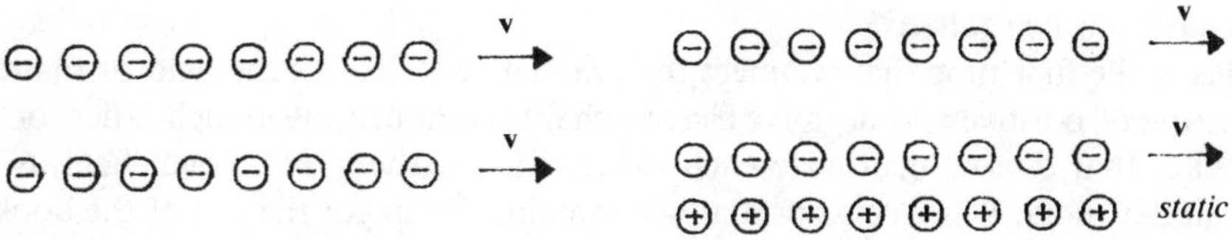

(a) (b)

Magnetic force and electrically neutralized charges

with the same velocity **v**. He will be observing two equal electric currents, though not wire-bound ones, flowing in the same direction. What he will see is given by (1) and (2) with $\mathbf{v}_1 = \mathbf{v}_2 = \mathbf{v}$; on resolving the double vector product, we have

$$\mathbf{F} = q(1 - \beta^2)\mathbf{E}_C \tag{4}$$

That is, he will see a strong electric force repelling the two rows of equal charges from each other, diminished very slightly by a β^2 times smaller magnetic force. I know of no experiment, nor can I imagine one in the foreseeable future, that could realize these conditions and demonstrate the slight decrease in repulsive force when two or more originally static charges are moved past an observer, without any charges of the opposite sign nearby.

To make this tiny force observable, we must first remove the electric force that overshadows it. The simple way to do this is to neutralize at least one of the rows by a row of stationary positive charges — stationary so as to keep a negative current flowing. That is, of course, essentially the case of a conductor, with the positive charges provided by the positive ion grid, and the electron flow forming the negative current. In other words, we are not able to demonstrate a magnetic force unless at least one of the currents flows in a normally neutral conductor such as a wire.

After pointing to this undisputed, but rarely mentioned circumstance, let me now discuss with respect to what the proposed theory defines the velocities $\mathbf{v}_1$ and $\mathbf{v}_2$ in (1) and (2). As always, it defines them with respect to the locally dominant force field. In the case of the electric force between two charges, this is of course the electric field of the other ("source") charge, through which the considered ("object") charge q is moving. But in the case of a magnetic force, this electric field must first be eliminated to make the magnetic force observable, so what is there left as the locally dominant field?

In the macroscopic world we live in, the positive charges, that is, the ion grid of the electrically neutral conductor, are at rest (or at best slowly moving) with respect to vast quantities of matter, which is likewise electrically neutral — the earth, in our neighborhood. The dominant force field, therefore, is the gravitational field through which the charge q in (1) and (2) is moving once the electric field has been neutralized.

This is the first time that we meet the gravitational field as the field of matter consisting of positive and negative electric charges that neutralize each other, or if you like, that *almost* neutralize each other, leaving the gravitational field as a "remainder" field. It is a concept that will remain with us for the rest of the book, and will be discussed in more detail in Part Three.

More important at the moment, however, is my hope that the case of the macroscopic magnetic force will make the idea of a rest standard given by the local gravitational field appear less exotic than it may have seemed at first.

To return to the Lorentz force (1) with the magnetic field defined by (2), the present theory defines the velocities occurring in them not with respect to an observer, as in the Einstein theory, but with respect to the dominant local field, which in the case of magnetic force due to wirebound currents is not the electric, but the gravitational field through which the charges are moving, as explained above.

From this it might be concluded that if the observer is at rest with respect to the gravitational field, which is usually the case, the two theories must be equivalent. Not so: we have already crossed the border from respectable orthodoxy into heresy. To see this, we first go back to some basic principles.

The Lorentz force (1) describes a force in terms of an interaction between a charge (q) and the field (**E**, for example) in its immediate environment. The charge that is the source of that field has not the slightest effect on this interaction; in fact, that charge may no longer exist (it could have been discharged before the collapse of its field arrived in the neighborhood of q). That is the first rule to remember if things get confusing.

Second, strictly speaking the fields such as **E** in (1) refer to *all* of the electric field in the immediate neighborhood of the charge q, including that charge's own field. However, it is clear from symmetry considerations that at rest or in uniform motion the net force on a charge by its own field is zero. It is therefore permissible, as is invariably done, to pretend that **E** is the field of the other charge only, and that q behaves as if it had no field. That is the second basic rule; but it applies only to uniform motion, including rest.

If the charge accelerates, partially catching up with its own field, the resulting asymmetry does produce a force on a charge by its own field. An easy way to see this is to recall that accelerating a charge means increasing its magnetic field, which will induce an electric field to oppose such a change, so that the charge is acted on by its own field. This "inertia of the electromagnetic field" will be discussed in the next section, and in greater detail in Part Two, but will not be needed here.

Both rules are still respectable physics, but once we define velocities with respect to the field, not with respect to an observer, we begin to obtain heretical results, as in the following example.

Let a charge q_2 be at rest with respect to the observer (gravitational field, ether), and let an equal and opposite object charge have velocity $\mathbf{v}$ in that rest frame. Then in the Einstein (and ether) theory $\mathbf{v}_2 = 0$, and there is no magnetic force. The moving charge does have a magnetic field, but in orthodox physics magnetic fields cannot affect stationary charges. Thus in orthodox electromagnetics we have

$$\mathbf{F} = q(\mathbf{E} + \mathbf{v} \times \mathbf{B}) = q\mathbf{E} \tag{5}$$

Now let the velocities be defined with respect to the local field through which the charge is traveling, or by the relativity principle, with respect to the field that is moving past the charge. Then $\mathbf{v}_2$ is the velocity of the charge with respect to the field in its vicinity, which is clearly the electric field of q_1 sweeping past it. Hence the velocity of q_2 with respect to the field surrounding it is $\mathbf{v}$, and (1) and (2), after resolving the double vector product as before, yield

$$\mathbf{F} = q\mathbf{E}(1 - \beta^2) \tag{6}$$

which we may interpret either as a diminished electric attraction, or as the appearance of a magnetic force. In either case, it is a result that differs from the prediction (5) of orthodox electromagnetics, and one that can, in principle, be subjected to an experimental test.

But only in principle. To test the difference, such an experiment would have to measure it with an accuracy of β^2 at the one moment when the velocity is perpendicular to the radius vector joining the two charges. On the other hand, if the electric field that swamps the effect is removed by neutralizing the stationary charge as discussed above, that is, by using a conductor with free electrons, (2) will indeed produce a zero magnetic field in the present theory, just as predicted by the Einstein theory, for the neutralized charge has no field with which to act, and the moving charge has nothing to act on. (Polarization effects etc. are not peculiar to *moving* charges.)

Now if both charges move, with the electric field neutralized by the presence of a stationary charge of opposite sign near one of them, or more generally, if we perform the experiment with a wire-bound current, then the force (1) between the moving charges is exactly what is predicted by the Biot-Savart Law – in all theories. That includes the present theory, for once the electric field has been neutralized, we fall back onto the remaining force field, which is gravitational, and which we also made the rest frame for the observer.

Thus, the original effect is too small to be measured, and trying to increase it will eliminate it. This is frustrating; but then, if the effect were easily measurable, the inaccuracy of orthodox electromagnetics would have been noted long ago.

The reader well versed in conventional electrodynamics may be offended by the idea that a stationary charge (stationary with respect to the observer) can be affected by a magnetic field. Yet what offends the relativity principle from my point of view is that a charge that is deflected when it travels through a magnetic field should sit still when a magnetic field sweeps past it. What possible difference can there be between the two cases?

None, of course, and the Einstein theory does ultimately obtain the same force in both cases. However, it does so by letting the observer see only an electric force in one case, and both an electric and magnetic force in the other, the two adding up to the same result. But to achieve this compliance with the relativity principle, space and time have to be distorted, and electric and magnetic fields have to be tied to the observer, instead of to the charges producing them.

By contrast, in the present theory electric and magnetic fields are quite independent of any observer, since the only velocity that is effective in producing a force is that of one charge in the field of the other. This relative velocity remains unchanged no matter what coordinate system an observer wishes to choose. The relativity principle is thus satisfied automatically and naturally, without time dilations or space contractions.

As for the magnetic field, it is uniquely determined by the electric field through (2), and we may express it via that electric field (as we have done above) to eliminate an unnecessary complication rather than convert parts of it to rescue the relativity principle for an observer who has nothing to do with the problem.

It will be seen that the entire behavior of a charge moving through a magnetic field is based on only two foundations 1) the charge experiences a magnetic force $q\mathbf{v} \times \mathbf{B}$, where $\mathbf{v}$ is referred to the *field*, 2) it is subject to the the relativity principle: its motion through a field cannot be distinguished from a field sweeping past the charge.

Although the basic principle is simple, things can become confusing when we have to find $\mathbf{B}$ from (2), and I will therefore suggest two "magnetic rules" that should simplify possible complications.

Magnetic rule no. 1 has, in effect already been derived: Whenever at least one set of moving charges has been electrically neutralized — most often by the positive ion grid of the conductor in which the charges are moving — then the remainder field with respect to which all velocities are defined is the gravitational field, which also coincides with the laboratory frame in all experiments on the point. Hence *in this case, which includes all cases of wirebound currents, Einsteinian electromagnetics remains valid.* Note that, as shown in our thought experiment above, it is sufficient to neutralize only one of the currents. For example, if we have an electron beam (with its electric field not neutralized) shooting through a magnetic field produced by wirebound currents in electrically neutral conductors, magnetic rule no. 1 will apply, so that mass, force, charge and everything else will obediently follow Einstein's predictions.

The difference between the present theory and orthodox electromagnetics arises only in interactions of charges whose electric field has *not* been neutralized. These are also the cases that have not been tested experimentally: the force between two parallel electron beams, or the force of one charge when moving in the field of another with no neutralizing charges present. The latter includes the case of planetary motion of the electron about the nucleus.

We need in principle consider only the case of two solitary charges, for the Maxwell equations are linear and therefore the case of an arbitrary charge distribution follows by superposition. As far as this book is concerned, we shall mainly need the case of planetary motion.

Since the velocity of a charge in a force field is defined as its velocity with respect to the lattice of intersections of lines of force and equipotentials, it may be convenient to think of the radial **E** field of a charge as drawn on the dial of a clock. The source charge is in the center of the clock, and 12 selected rays are labeled "one o'clock" to "noon." When one charge moves through the dial (field) of the other, we can then assign the sign of the velocity depending on whether the motion is clockwise or anticlockwise. (A moving charge has no magnetic field in the longitudinal direction, so radial motion is irrelevant.)

The use of this method is trivial when the motion of the charge in the field of the other is rectilinear; however, for planetary, circular motion of one charge about the other, we must remember that rotation of a charge about its own axis (such as electron spin or the earth's rotation) does not affect the field: we know from the Michelson-Gale [1925] experiment (Sec. 1.3.7) that the terrestrial gravitational field is no more affected by the earth's rotation than is the plane of Foucault's pendulum. Analogously, we assume any electron spin to be irrelevant to its **E** field. Therefore an electron moving clockwise through the nucleus dial from 2 to 4 o'clock will simultaneously have its field sweep past the nucleus with its 8 to 10

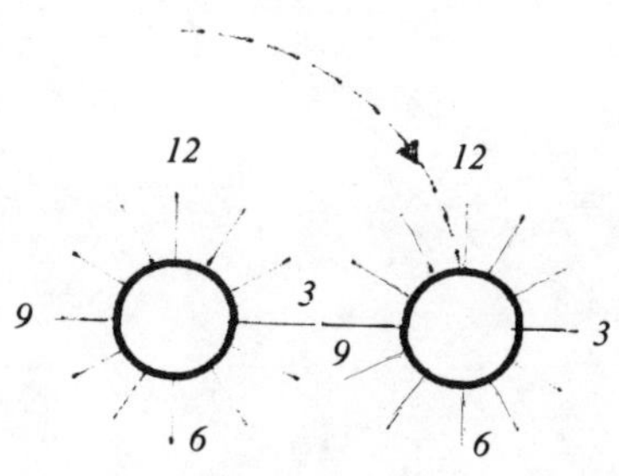

Charge through field and field past charge

o'clock rays, and a quarter of a period later with its 11 to 1 o'clock rays. Hence the velocity of the electron in the field of the nucleus is the same in sign and absolute value as the velocity with which the field of the electron sweeps past the nucleus.

Thus, setting $\mathbf{v}_1 = \mathbf{v}_2 = \mathbf{v}$, we find magnetic rule no. 2: *The force between two solitary charges, none of which is electrically neutralized by other charges, is*

$$\mathbf{F} = q(\mathbf{E} + \mathbf{v}\times\mathbf{B}) = q(\mathbf{E} - \beta^2\mathbf{E}_c) \tag{7}$$

where $\mathbf{E}_c$ is the Coulomb field, i.e., the irrotational part of $\mathbf{E}$. The rotational part of $\mathbf{E}$, the Faraday field, is absent if there is no acceleration, and has no effect on the force between the two charges when the acceleration is perpendicular to the velocity (as it is in circular planetary motion). In both of these cases (7) reduces to

$$\mathbf{F} = q\mathbf{E}(1 - \beta^2) \tag{8}$$

The rule expressed by (7) or (8) amounts to this: the magnetic field of a moving electric charge, being uniquely determined by (2), can be eliminated as an unnecessary encumbrance. The resulting modification of the electric field is always such as to *diminish* the electric force (this is equally valid for repulsion).

The magnetic force between two moving charges is, of course, the Biot-Savart force. It appears here as a simple modification of the Coulomb force and, as always in the present theory, as a force independent of the observer. This differs from the Einstein theory in a result that is not yet susceptible to measurement. An Einsteinian observer located at the center of mass in circular planetary motion will calculate a Biot-Savart force that is of order β^3 and acts in the opposite direction.

1.5. Electromagnetic momentum

When hammer hits chisel, the elastic steel of the chisel transmits the force from the hammer to the cutting edge of the chisel. In the same way, Faraday and Maxwell thought, the ether transmitted the force from one point charge to another, and Maxwell calculated the force per unit area (that is, the stress, such as pressure or tensile stress) exerted by a charge on a surface in the ether. He did so in much the same way that one would calculate the stress on the cross section of a rod.

The concept of such an elastic ether has been abandoned, but since we have no reason to doubt the Maxwell equations (at low velocities, anyway), these calculations remain formally correct, for they are based on nothing else. The word "stress" is generally associated with forces between the particles of a material medium, but though such stresses do occur in ponderable materials permeated by electromagnetic fields, there is obviously nothing under stress in a vacuum. This inappropriate word has remained in use, but in the present theory, which regards force as propagating with velocity c through the local force field, it might better be called the "force density" (in N/m^2), meaning the force transmitted through unit area perpendicular to its flow.

To calculate it, we start from the Maxwell equations for a region which contains charge and current distributions, but consists otherwise of vacuum:

$$\nabla \times \mathbf{E} + \frac{\partial \mathbf{B}}{\partial t} = 0 \tag{1}$$

$$\nabla \times \mathbf{B} - \frac{1}{c^2}\frac{\partial \mathbf{E}}{\partial t} = \mu_0 \mathbf{J} \tag{2}$$

$$\nabla \cdot \mathbf{E} = \frac{\rho}{\epsilon_0} \tag{3}$$

$$\nabla \cdot \mathbf{B} = 0 \tag{4}$$

Cross-multiplying (1) by $\epsilon_0 \mathbf{E}$, and (2) by $\mathbf{B}$, then adding and rearranging using $1/\epsilon_0\mu_0 = c^2$, we obtain

$$\epsilon_0(\nabla \times \mathbf{E}) \times \mathbf{E} + \frac{1}{\mu_0}(\nabla \times \mathbf{B}) \times \mathbf{B} = \mathbf{J} \times \mathbf{B} + \epsilon_0 \frac{\partial}{\partial t}(\mathbf{E} \times \mathbf{B}) \tag{5}$$

By a simple, but somewhat longwinded calculation using (3) and (4) and given in many textbooks (e.g., [Stratton 1941]), the first two terms can be shown to equal

$$\begin{aligned} \epsilon_0(\nabla \times \mathbf{E}) \times \mathbf{E} + \frac{1}{\mu_0}(\nabla \times \mathbf{B}) \times \mathbf{B} &= \operatorname{div} {}^2\mathbf{S} - \epsilon_0 \mathbf{E}\, \nabla \cdot \mathbf{E} - \frac{1}{\mu_0}\mathbf{B}\, \nabla \cdot \mathbf{B} \\ &= \operatorname{div} {}^2\mathbf{S} - \rho \mathbf{E} \end{aligned} \tag{6}$$

where ${}^2\mathbf{S}$ is the electromagnetic stress tensor. Its components are of no interest here; however, it is evident that ${}^2\mathbf{S}$ is a quantity that, when integrated over a closed surface Σ, must yield the total force transmitted across it. Substituting (6) in (5), using $\mathbf{J} = \varrho\mathbf{v}$, $\int \varrho\, dV = q$, and integrating (5) over a volume V, we therefore have by the Divergence Theorem

$$\mathbf{F} = q\mathbf{E} + q\mathbf{v} \times \mathbf{B} + \frac{1}{c^2}\frac{\partial}{\partial t}\int \mathbf{E} \times \mathbf{H}\, dV \tag{7}$$

The first two terms on the right are obviously the electric and magnetic components of the Lorentz force by which the corresponding fields act on a charge, and this is, in fact, how the Lorentz force is derived from the Maxwell equations.

But (7) also says that even if there are no charges or currents in the considered volume, so that the first two terms are zero, there is still a net force emanating from this "empty" volume, whenever it is permeated by a time-varying electromagnetic field. Since force is the rate of change of momentum, it follows that a momentum

$$\mathbf{p} = \int \frac{\mathbf{E} \times \mathbf{H}}{c^2}\, dV \tag{8}$$

must be associated with an electromagnetic field. This phenomenon is sometimes called "inertia of the electromagnetic field."

The physical meaning is the following. The momentum of an *uncharged* body $m\mathbf{v}$, when changed by external forces, seeks to stay constant and resists such a change. But so does an electromagnetic field, and quite independently of the momentum and inertia of the mechanical, Newtonian masses that carry its source charges and currents. A steady magnetic field, for example, is due to a steady current; if that field is changed (by changing the current) it will, by Faraday's Law, induce an electric field that will seek to restore the current and its magnetic field to its previous value — its direction is given by Lenz's Law, and the entire effect is known as self-inductance (mutual inductance if the field was changed by another current). The magnetic field, in effect, resists being changed.

Quite similarly, a steady electric field is due to a steady charge distribution. If the field is changed (by moving the charges), the resulting displacement current $\partial\mathbf{D}/\partial t$ gives rise to a magnetic field $\mathbf{B}$ by (2), and the change in magnetic field induces an $\mathbf{E}$ directed against the displacement of the charges. The electric field, in effect, resists being changed.

But if the momentum of a field parallels the momentum of uncharged matter, we would expect an inertial mass of the field to parallel the inertial Newtonian or mechanical mass of an uncharged body. This is indeed the case: we shall, in a moment, find the electromagnetic mass of a charged body as the factor multiplying $\mathbf{v}$ in the expression for the momentum ($m\mathbf{v}$) of an electromagnetic field. In both cases, mechanical and electromagnetic, inertial mass is a measure of a body's resistance to having its momentum changed.

This electromagnetic mass is no formal mathematical trick. It is a physical reality that a charged body resists acceleration *beyond* the resistance offered to it by its Newtonian mass.

To see that the inertial mass of an uncharged body is increased by an *additional* electromagnetic mass of its field when that body is given a charge, consider an example that will be used several times in coming sections, the throwing of a tennis ball. When it is uncharged, its Newtonian (mechanical) inertial mass resists acceleration, and the work done by the thrower's muscles in overcoming that resistance appears as kinetic energy of the moving ball. But when the ball is electrically charged, the ball offers additional resistance (in principle, that is, for the numerical amount is actually very small): a moving charge has a magnetic field proportional to its velocity, and the change in magnetic field (from zero), by Faraday's Law, induces an electric field opposing the acceleration of the charge. The additional muscle work performed in overcoming this resistance appears as the energy of the magnetic field in addition to the ball's kinetic energy.

Now let us calculate this electromagnetic mass of a body as the factor multiplying its velocity to yield its electromagnetic momentum. We consider a moving point charge, which by the Divergence Theorem is also equivalent to any spherical charge distribution with radial symmetry; by superposition, we may consider *all* (reasonable) charge distributions made up of such elementary spherical charges.

Substituting for $\mathbf{H} = \mathbf{B}/\mu$ from (2), Sec. 1.4, in (8), resolving the double cross-product, directing the x-axis along $\mathbf{v}$, and omitting the terms that will integrate to zero because of symmetry considerations, we find the momentum in the form

$$\mathbf{p} = \frac{\epsilon \mathbf{v}}{c^2} \int (E_y^2 + E_z^2)\, dV \tag{9}$$

Hence the electromagnetic or field mass, the factor multiplying $\mathbf{v}$, is

$$m_f = \frac{\epsilon}{c^2} \int (E_y^2 + E_z^2)\, dV \tag{10}$$

Let us now temporarily assume that the Coulomb field remains spherically symmetrical when its source charge moves with respect to the rest frame (by letting the observer rest in the local force field or the ether, we need not yet differentiate between the three theories). In that case the three rectangular components of $\mathbf{E}$ will remain equal, as they were at rest, and we have

$$E_y^2 + E_z^2 = \tfrac{2}{3} E^2 = \frac{2K^2}{3r^4 q^2} \tag{11}$$

where the constant K is defined in (2), Sec. 1.1. Substituting this in (10) and integrating over all space outside the charge (which we assume distributed over the surface of a sphere with radius R), we find the electromagnetic mass

$$m_f = \frac{\mu q^2}{6\pi R} \tag{12}$$

This is a result that we will also obtain in Part Two by several other methods. However, it is valid only to the extent that the assumption of field symmetry, on which (11) is based, is valid. This is evidently the case for slow (uniform) velocities, which must merge continuously with the static case.

But for high velocities, the story is different. There is no direct experimental evidence available, and we must trust the Maxwell equations to provide the answer. To evaluate the expression for the electromagnetic mass (10) exactly, we must first examine what happens to the field of a fast moving point charge.

1.6. The Field of a Moving Charge

We will now consider a seemingly simple problem. We take a point charge at rest with its concentric equipotential spheres and radial Coulomb field. What happens to this potential ϕ and electric field $\mathbf{E}$ when the charge moves with uniform velocity $\mathbf{v}$ (directed along the x-axis) with respect to the rest frame? As before, we let the observer sit still in the local force field or in the ether for comparison; but we measure velocities with respect to the local force field in which the charge is moving.

We have but two tools to solve this simple problem: the Maxwell equations and the relativity principle.

The field vectors $\mathbf{E}$ and $\mathbf{B}$ satisfying the Maxwell equations are derivable from a scalar potential ϕ and a vector potential $\mathbf{A}$. As shown in textbooks of electromagnetism (and also in Part Two of this book), the relations are

$$\mathbf{E} = -\frac{\partial \mathbf{A}}{\partial t} - \nabla\phi \tag{1}$$

$$\mathbf{B} = \nabla \times \mathbf{A}; \qquad \nabla \cdot \mathbf{A} = -\frac{1}{c^2}\frac{\partial \phi}{\partial t} \tag{2}$$

where the potentials are solutions of the wave equations

$$\nabla^2\phi - \frac{1}{c^2}\frac{\partial^2\phi}{\partial t^2} = -\frac{\rho}{\epsilon_0} \tag{3}$$

$$\nabla^2\mathbf{A} - \frac{1}{c^2}\frac{\partial^2\mathbf{A}}{\partial t^2} = \mu_0\rho v \tag{4}$$

Next, we turn to the principle of relativity, which requires that the laws of physics, when properly formulated, remain equally valid in all frames moving with uniform velocity with respect to each other. That means whatever the field distribution about a point charge, however it is affected by the particle's velocity, and whatever that velocity is referred to, the field distribution must travel unaltered with the particle ("frozen to it") whenever it moves with uniform velocity: otherwise we could – in principle, anyway – look at the distortion of the field surrounding the particle, and without reference to any rest standard, we could proclaim with what absolute velocity the particle is moving. The principle of relativity therefore requires the "freeze" condition: as the charge moves through the rest frame with velocity $\mathbf{v}$, each component of its field must satisfy the relation

$$f(x, y, z, t) = f(x + v_x dt,\ y + v_y dt,\ z + v_z dt,\ t + dt) \tag{5}$$

From this we have

$$\frac{\partial f}{\partial t} = -v_x\frac{\partial f}{\partial x} - v_y\frac{\partial f}{\partial y} - v_z\frac{\partial f}{\partial z} = -\mathbf{v}\cdot\nabla f \tag{6}$$

and using this relation twice over to eliminate the time derivative in (3), we find

$$(1-\beta^2)\frac{\partial^2\phi}{\partial x^2}+\frac{\partial^2\phi}{\partial y^2}+\frac{\partial^2\phi}{\partial z^2}=-\frac{\rho}{\epsilon_0} \tag{7}$$

where

$$\beta=\frac{v}{c} \tag{8}$$

This is certainly different from the Poisson equation when the charge is at rest, due to the $(1-\beta^2)$ factor. Yet (7) is as valid as the relativity principle and the Maxwell equations.

There have been several attempts to interpret this result in a way that is consistent with the experimental evidence without sacrificing either the relativity principle or the Maxwell equations, both of which underlie (7). Not all of these attempts have been successful.

Here I would like to insert parenthetically that while I would not like to sacrifice the relativity principle, I lack the obligatory reverence toward the Maxwell equations: they are ether-begotten and tested without circularity only at low velocities. There are, however, two reasons why I am not ready to abandon them. The first is obvious: I have nothing better to offer. The other will be discussed in Sec. 1.10.1, which explains the fundamental reason why the Maxwell equations can very well survive without the elastic ether of which they were born.

The two important methods of interpreting (7) are due, respectively, to Hendrik Lorentz and Albert Einstein.

Lorentz, who took $\mathbf{v}$ to mean the velocity with respect to the ether, noted that (7) is equivalent to the electrostatic case when the charge is at rest, provided x is replaced by x', where

$$x'=\frac{x}{\sqrt{1-x^2}} \tag{9}$$

In that case (7) turns into

$$\nabla^2\phi=-\frac{\rho}{\epsilon_0} \tag{10}$$

where

$$\rho=\frac{1}{\sqrt{1-\beta^2}}\;\rho_0\left(\frac{x}{\sqrt{1-\beta^2}},y,z\right) \tag{11}$$

with ϱ_0 the rest charge density.

From this Lorentz concluded that electrons contract in the direction in which they move through the ether. It was for this purpose that he introduced the transformation named after him, a very different purpose from that for which Einstein used it. (This, incidentally, also explains the seeming paradox why the man who provided the backbone of the Einstein theory remained irreconcilably opposed to it to his death in 1928.)

But the experimental evidence decided against the Lorentz theory. As the earth moved through the (unentrained) ether, a measurable torque would act on two charges attracting each other in the "ether wind" (as will be explained in Sec. 1.10.4), but the Trouton-Noble experiment [1903] failed to detect such a torque.

It is, however, important to note that Lorentz's theory (or Ritz's, for that matter), did not contradict the relativity principle; it contradicted the experimental evidence. The contraction of an electron (or in Fitzgerald's view, of *any* body) as it moves through an ether is no more violation of relativity than rain drops being deformed as they fall through the atmosphere; the fact that in reality there is no such contraction has as little to do with the relativity principle as the invalid theory that sailboats always move against the wind.

To Einstein, on the other hand, $\mathbf{v}$ was the velocity with respect to the observer; if he wanted all observers to see the same force between two charges, he had to build the contraction (9) into space itself, which in turn made the dilation of time inevitable. This was consistent with the Maxwell equations using observer-referred velocities and squared with the experimental evidence, for there was no longer an ether to be refuted. Whether it satisfied the relativity principle, however, is a point less self-evident than the textbooks would have us believe. It certainly did not satisfy it for a world with undistorted space and time, nor for one that believed nature's laws, including the relativity principle itself, to be rooted in nature, not in the observer observing it. Perhaps it is fair to say that the Einstein theory satisfies the relativity principle at a price that not everybody is willing to pay.

In the present theory, $\mathbf{v}$ is the velocity with respect to the local force field. There is no privileged standard of rest, and since the effect-producing velocity in Maxwell's equations and in the Lorentz force is held to be the vector difference between the velocities of a charge and the local force field as both move through an arbitrary observer's coordinate system, this effective velocity is unchanged and independent of any observer. It therefore satisfies the relativity principle automatically. (More on this in Sec. 1.10.4.)

Nor is there any reason to abandon the Maxwell equations, which likewise become observer-independent (more on this, again, in Sec. 1.10.4). Judged by the present theory, the Trouton-Noble experiment (Sec. 1.10.4) produced a null effect because nothing moved. The two charges were at rest with respect to each other, i.e., at rest in each other's fields.

To return to the basic problem, is there a good physical reason why a uniform charge on a sphere at rest should rearrange itself as implied by (11) when the charge moves?

Einstein did not need one; in effect, he regarded the Maxwell equations, with all velocities referred to the observer, as an axiom and distorted space and time to fit them. Lorentz, who regarded the ether "as endowed with a certain degree of substantiality, however different it may be from all ordinary matter" [Lorentz 1915], followed Maxwell's thinking of actual stresses in the ether and thought of the electron contraction as the result of stresses by the ether on a deformable electron.

The present theory offers an entirely different physical explanation – but not at this point, for we need certain fundamental relations that will not be derived until Part Two. For the time being, therefore, the reader is asked to accept (11) for no better reason than that it emerges from the Maxwell equations – a procedure that should present Einsteinians with no difficulty at all.

Now let us turn to the field produced by such a non-uniform charge distribution on a moving sphere. We can solve (7) by *pretending* that we are dealing with an electrostatic problem in a space in which the x coordinates have contracted in accordance with (9) – another just-as-if equivalence, and one that the Einstein theory regards as a physical reality. The charge distribution (11) will then lead to a potential

$$\phi = \frac{1}{\sqrt{1-\beta^2}}\,\phi_0\left(\frac{x}{\sqrt{1-\beta^2}}, y, z\right) \tag{12}$$

In the special case of a point charge, or its equivalent, a sphere with radially symmetrical charge distribution, this becomes

$$\phi = -\frac{q}{4\pi\epsilon_0}\frac{1}{\sqrt{x^2+(1-\beta^2)(y^2+z^2)}} = \frac{\phi_0}{\sqrt{1-\beta^2\sin^2\theta}} \tag{13}$$

where θ is the angle between the radius vector and the velocity.

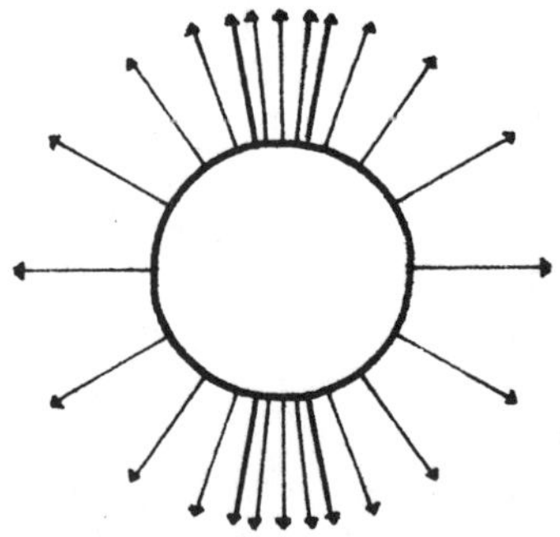

The "bunching" of the electric field in the direction perpendicular to the velocity of a charge. In the Einstein theory, this is due to the contraction of length in the direction of the velocity referred to an *observer*; in the present theory, it is due to a redistribution of the charge density determined by the velocity with respect to the *traversed field*.

Thus the familiar concentric spherical equipotentials about a static point charge flatten into ellipsoids when the charge moves with high velocity. The Coulomb field is perpendicular to the equipotentials, so that if we take the velocity as the north-south axis of a sphere, the electric lines of force are less dense near the poles and denser near the equator – we shall refer to this effect as "bunching" of the lines of force. To calculate it, we take the gradient of (13), obtaining

$$\mathbf{E} = \frac{E_0(1-\beta^2)}{(1-\beta^2 \sin^2 \theta)^{3/2}} \mathbf{r}_0 \tag{14}$$

This result is often needed in terms of the field parallel and perpendicular to the velocity:

$$E = \begin{cases} E_0(1-\beta^2) & \text{for } \mathbf{r} \parallel \mathbf{v} \\ \dfrac{E_0}{\sqrt{1-\beta^2}} & \text{for } \mathbf{r} \perp \mathbf{v} \end{cases} \tag{15}$$

Note that (11) implies only a rearrangement, not a change, of the total charge, which remains conserved:

$$\begin{aligned} \iiint \rho \, dV &= \iiint \rho_o \left(\frac{x}{\sqrt{1-\beta^2}}, y, z \right) \frac{dx}{\sqrt{1-\beta^2}}, dy, dz \\ &= \iiint \rho_0 \, dV = q \end{aligned} \tag{16}$$

The reader is reminded that equations (7) through (16), though not the text in-between, are also formally valid in Einsteinian electromagnetics. There are, however, two substantial differences. First, $\mathbf{v}$ refers to the observer in the Einstein theory, but to the local force field in ours; second, the contraction (9) applies to space itself in the Einstein theory, but only to the charge density (and consequently to the field it produces) in ours.

1.7. Mass and Energy

With the electric field strength of a moving charge established, we are now ready to substitute in the expression for electromagnetic mass (10), Sec. 1.5.

We recall that the velocity was parallel to the x-axis; to substitute for E_y and E_z, we use (15), Sec. 1.6; then (10), Sec. 1.5, becomes

$$m_f = \frac{\epsilon}{c^2}\frac{1}{\sqrt{1-\beta^2}}\int\left[E_{0y}^2\left(\frac{x}{\sqrt{1-\beta^2}}, y, z\right)\right.$$
$$\left. + E_{0z}^2\left(\frac{x}{\sqrt{1-\beta^2}}, y, z\right)\right] dx\,dy\,dz \tag{1}$$

On substituting $\xi = x/\sqrt{(1-\beta^2)}$, this integrates to

$$m_f = \frac{\mu q^2}{6\pi R\sqrt{1-\beta^2}} \tag{2}$$

or using (12), Sec. 1.4,

$$m_f = \frac{m_{f0}}{\sqrt{1-\beta^2}} \tag{3}$$

where m_{f0} is the electromagnetic mass at rest.

This formula for the electromagnetic mass, associated with the resistance of the electromagnetic field to the acceleration of its source charge, had been known to the classics of the late 19th century before the advent of the Einstein theory, which derives the same type of formula for *any* mass, charged or neutral.

The resistance offered to change of momentum by *uncharged* matter evidently obeys the same law. First and foremost, this is supported by the experimental evidence.

But there are also theoretical reasons. It was shown by Page [1912] that Coulomb's Law plus the Lorentz transformation is enough to derive the Maxwell equations, and hence all the experimental evidence confirming the Einstein theory. But Coulomb's Law is formally identical with Newton's inverse square law, though its numerical values and dimensions may differ, and Page's method must therefore lead to formally identical results. Hence (3), which rests on nothing but the Maxwell equations and the relativity principle, must equally well hold for Newtonian, mechanical, electrically neutral mass.

By this Page-Coulomb-Lorentz argument (and, of course, the experimental evidence), we may therefore write

$$m_n = \frac{m_{n0}}{\sqrt{1-\beta^2}} \tag{4}$$

where m_{n0} is the neutral mass at rest.

Since the total inertial reaction to an impressed force acting on a charged body moving with velocity $\mathbf{v}$ is

$$\frac{d}{dt}(m_n\mathbf{v} + m_f\mathbf{v}) = \frac{d}{dt}(m_n + m_f)\mathbf{v} \tag{5}$$

we can combine the two masses into a single mass

$$m = m_n + m_f \tag{6}$$

without knowing what fraction is made up of one kind or the other. Combining (3) and (4) in this way yields

$$m = \frac{m_0}{\sqrt{1-\beta^2}} \tag{7}$$

The classics were aware of (3), but unsure of (4); they hoped that Newtonian mass would turn out to be constant, so that the two components could eventually be separated by advancing technology, i.e., by accelerating particles to sufficiently high velocities. Only Newton himself, who did not even know of the existence of electromagnetics, had guarded against the possibility of mass being velocity-dependent by never taking it out of the momentum entity when dealing with force.

Contemporary physics claims that the resolution of mass into its two components (6) is unimportant, or even impossible. This is incorrect: in Part Two the resolution will be shown important and possible.

Equation (7), perhaps due to the abstract and overly mathematical character of the Einstein theory, has been imbued with a mystic romanticism which it does not deserve. The quantity of matter does not, of course, increase with velocity. What increases is the inertial reaction or the resistance to a force changing a body's momentum. But that is nothing extraordinary: inertial reaction is a force, and there are many forces that are velocity-dependent — hydraulic or aerodynamic friction, for example, and the thrust by a ship or plane to overcome them.

Other formulas widely used in the experimental verification of the Einstein theory follow from (7), and like (7) itself, they can again be derived without use of the Lorentz transformation.

Writing (7) in the form

$$m^2(1-\beta^2) = \text{const} \tag{8}$$

and differentiating, we have after elementary manipulations

$$v^2\,dm + vm\,dv = c^2\,dm \tag{9}$$

On the other hand, the work done, or energy (E) expended, in accelerating a given quantity of matter from velocity 0 to velocity v is

$$E_{kin} = \int_0^v \frac{d(mv)}{dt}\, ds = \int_o^v v\, d(mv) = \int_0^v [v^2\, dm + vm\, dv] \tag{10}$$

Substituting (9) for the integrand, this yields the kinetic energy

$$\begin{aligned} E_{kin} &= c^2 \int_{m(0)}^{m(v)} dm = m_0 c^2 [m(v) - m(0)] \\ &= m_0 c^2 \left(\frac{1}{\sqrt{1-\beta^2}} - 1 \right) \end{aligned} \tag{11}$$

which, as is easily seen by series expansion, reduces to the familiar $\frac{1}{2}mv^2$ for small β. However, (11) shows that this kinetic energy is merely the difference of energies at velocity v and velocity 0, the latter indicating an energy associated with a body at rest and given by the zero-velocity terms in (11):

$$E_0 = m_0 c^2 \tag{12}$$

The total energy of matter moving with velocity v is therefore

$$E = E_{kin} + E_0 = mc^2 \tag{13}$$

Finally, to establish a relation between the energy of a moving particle and its momentum, we have from (13)

$$\frac{E^2}{c^2} - m_0^2 c^2 = \frac{m_0^2}{1-\beta^2} - m_0 c^2 = \frac{m_0^2 v^2}{1-\beta^2} \tag{14}$$

The last expression is clearly the square of the momentum $\mathbf{p} = m\mathbf{v}$; therefore the required relation is

$$\frac{E^2}{c^2} = m_0 c^2 + p^2 \tag{15}$$

The relations that have been used most often in alleged proofs of the Einstein theory are (7), (13) and (15), as we shall see in more detail in Sec.1.9. None of them, as shown here, need the Lorentz transformation or the reformation of space and time.

Relation (13) has fascinated laymen, for it is often the only thing they know about the Einstein theory. But even some physics professors have romanticized "the equivalence of mass and energy." A glance at (13) shows that this "equivalence" is a dimensional absurdity.

The interpretation of (13) is rather simple. We have seen that the electromagnetic field resists acceleration of the charge that is its source, so that it is responsible for part of the inertial mass of the body carrying the charge. If we cause the charge to disappear by discharging the body, the field disappears only from its immediate surroundings: it is radiated away. But the energy of the field is radiated away with it, so that the conservation of energy demands that the inertial mass of the body be decreased by a corresponding amount. I see no reason to doubt that the same is true of the Newtonian part of the inertial mass, though we cannot demonstrate it by "dismassing" a body as we can discharge it, at least not in the macroscopic world.

This is the interpretation Einstein gave the relation (13); in fact, I know of no simpler way to express it than Einstein himself did in a special paper devoted to the point [1905b]. Emphasized by his own italics, Einstein's statement is *If a body gives off the energy L in the form of radiation, its mass diminishes by L/c^2.*

The classics came very close to deriving (13); in fact, it has been claimed that they were well aware of it. In 1900, for example, Henri Poincaré calculated the recoil experienced by a body radiating an energy E and found it by equating it to the momentum of the radiated electromagnetic field, given by (8), Sec. 1.5. This led him to a formula implying that the mass M associated with the radiated field equals E/c^2; however, this is not equivalent to Einstein's formula, as it deals only with the mass equivalent in radiation pressure, not with the mass lost as radiated energy. He may, however, have come closer in [Poincaré 1904]. This and some other sources are discussed by Ives [1952], who claims that Einstein's derivation is neither original nor correct. However, Ives' vehement animosity toward Einstein may have driven him too far here, and the reader interested in the history of the relation is cautioned to look up the original sources quoted by Ives before accepting his interpretation of what they imply.

1.8. The Modified Newton-Coulomb Law

Substituting the velocity dependent mass (7), Sec. 1.7, in Newton's Second Law in the only form Newton ever stated and used it, we have

$$\mathbf{F} = \frac{d}{dt}(m\mathbf{v}) = \frac{m_0}{(1-\beta^2)^{3/2}}\frac{dv}{dt}\mathbf{v}_0 + \frac{m_0}{\sqrt{1-\beta^2}}\frac{dv}{dt}\mathbf{u}_0 \tag{1}$$

where $\mathbf{v}_0$ is a unit vector in the direction of the velocity, and $\mathbf{u}_0$ is a unit vector at right angles to it. The two terms in this formula, which formally agrees with the formula for force in the Einstein theory, correspond to the components of acceleration directed along the velocity and perpendicular to it, respectively.

Thus, when the acceleration is normal, or close to normal, to the velocity (the magnetic force is *always* normal to it), then on comparing the second term to the second line in (15), Sec. 1.6, we have

$$\mathbf{F}_\perp = q(\mathbf{E}_{0\perp} + \mathbf{v}\times\mathbf{B}) = m_0\frac{dv}{dt}\mathbf{u}_0 \tag{2}$$

That is, not only does the Lorentz force remain valid, but the inertial reaction to it is given by the "mass times acceleration" formula falsely attributed to Newton. In this "transversal" case it is correct.

The formula is formally correct in the Einstein theory, too. Though this is usually known only to the more erudite Einstein scholars, the non-Newton formula "force = mass times acceleration" is perfectly valid in the Einstein theory for "transversal mass," that is, for the effective inertial mass when the acceleration is normal to the velocity – as can be seen from (1), which is obviously valid in the Einstein theory, too. But the agreement between these expressions and those in the Einstein theory is only formal, for Einstein defines "rest" with respect to the observer, not the field. Therefore the field $\mathbf{E}$ has a different value for different observers. It is, in fact, well worth to take a little side trip into the Einstein theory to see what enormous complications hide behind its seemingly simple formula

$$\mathbf{F} = q(\mathbf{E} + \mathbf{v}\times\mathbf{B}) = \frac{d}{dt}(m\mathbf{v}) \tag{3}$$

even in the elementary case of two equal and opposite charges attracting each other when one is in motion. Let the charge *2* be at the origin of the observer's system at time $t=0$, and let the moving charge *1* move at right angles to the radius vector at that moment. Moreover, let the charge at the origin be so massive that the force acting on it makes it move only very slowly so that its mass M can be considered equal to its rest mass at all times, and the magnetic force is, in the Einstein theory, quite insignificant. Then on using (15), Sec. 1.6, the force of the moving charge *1* with its "bunched up" field on the stationary charge *2* is

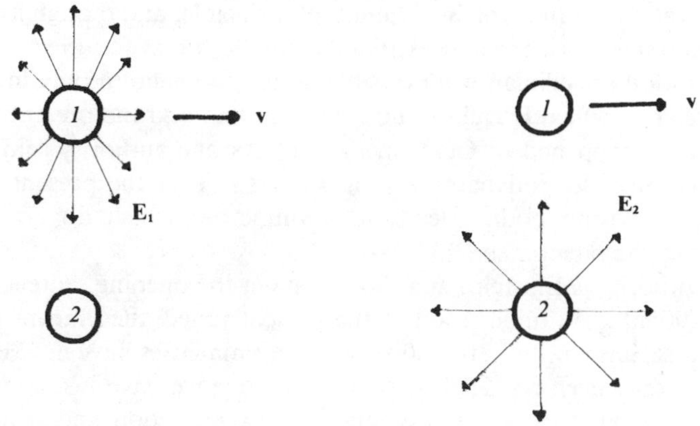

(a) *(b)*

Inequality of action and reaction in the Einstein Theory: (a) force of a moving charge on a stationary one, (b) force of a stationary charge on a moving one.

$$\mathbf{F}_{12} = q\mathbf{E}_1 = \frac{q\mathbf{E}_0}{\sqrt{1-\beta^2}} \tag{4}$$

However, the force exerted by the rest field of charge *2* on the moving charge *1* is

$$\mathbf{F}_{21} = q\mathbf{E}_2 = q\mathbf{E}_0 \tag{5}$$

which differs from (4) by a factor of up to infinity.

Action and reaction are thus no longer equal. (Note that we have never left the observer's system, so that this cannot be blamed onto a faulty transformation.)

Once again, it takes an erudite Einsteinian to resolve this paradox appearing in this simplest of all problems in the domain of moving charges. The explanation: There is no paradox. It is nowhere written that the two forces must be equal.

Yes, it is, one might think. If they are not equal, then the customary derivation of the conservation of momentum breaks down. Well, yes and no, say the Einsteinians. If the two bodies exerting a force on each other are in actual contact, as they are in the collision of billiard balls, then we know where here and now is, and conservation of momentum reduces to the classical meaning. But if we have action at a distance, then simultaneity is a concept that slides along slippery "world lines," and the where, when and how of a momentum, at least some of which is associated with a field stretching from here-now to four-dimensional eternity, becomes a somewhat nebulous concept. The conservation of momentum can be treated only in a generalized form involving Hamilton's principle, and though its conservation, in a certain sense, is ultimately extruded from the goo of four-vectors, the less than highly erudite Einsteinian will do better to use momentum only in the context of (15), Sec. 1.7, which is valid in both the Einstein and the present theory.

Whatever happened to Ockham's razor? As one authority said, "It is known that Maxwell's electrodynamics – as understood at the present time – when applied to moving bodies, leads to asymmetries which do not appear to be inherent in the phenomena."

The authority is Einstein, and the statement the opening sentence of his classic paper [1905a]. "As understood at the present time," then meant the ether, and now means Einstein; but after 80 years the asymmetries have not been eliminated, they have only been replaced by such as the one we have just considered.

In the present theory, the asymmetry between action and reaction *has* been eliminated. The two forces are not only equal (and opposite), but utterly indistinguishable: there is no way of telling which of the two charges is "genuinely" moving. In the Einstein theory, this is not possible, either; but this impossibility is tied to that of observing the simultaneity of two events that are not also coincident in space.

And that brings us to the crux of the matter: the Einstein theory (and all of contemporary physics) deals in *observables*. But as the case of the railroad track in the Introduction shows, it is better to deal in *inferrables*, and that is exactly what the present theory does.

For example, it is true that none of us know what happened to the location of the sun in the last 8 minutes (which is roughly the time it takes for its light to reach us); but we have a pretty good idea where it is right now nevertheless, for we can use the laws of nature to infer its present position, even though we cannot possibly measure it directly.

The planets orbiting the sun, and the electrons orbiting the nucleus, are, in fact, good examples of applying the present theory, for this will enable us to generalize the static Newton-Coulomb Law

$$\mathbf{F} = \frac{K}{r^2}\mathbf{r}_0 \tag{6}$$

to the case when the two masses or charges are in motion – one in the field of the other.

Consider the following problem: Is the bright disk in the sky a souvenir left by the sun where it was 8 minutes ago, or is that the direction to the real sun where it is now?

Let the sun S be at the origin of the coordinate system and let the earth be at E_1 at a distance r, and let it move along a circular orbit with orbital (circumferential) velocity v. Then during the time $t=r/c$ that it takes for the light to propagate from sun to earth, the latter will have moved through a distance $vt=\beta r$, where $\beta=v/c$. That is, it will have moved through an angle β. Note that this angle is independent of the distance r, which might as well be infinite, so that in determining the direction of the true sun we need consider only angles, not distances; in particular, parallel lines in our distorted figure can be considered identical. Now when the earth moves through the arriving light with its own

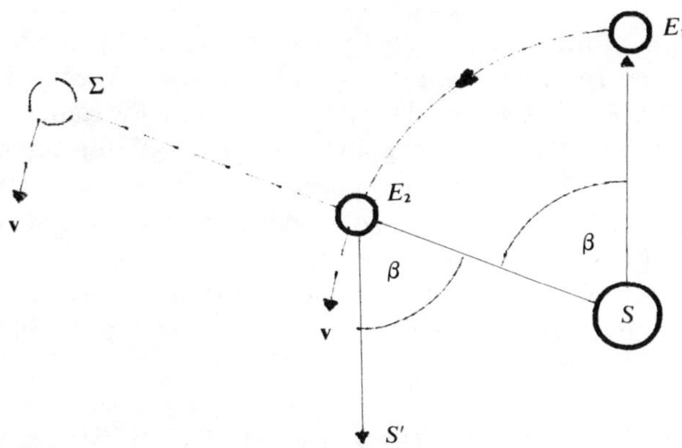

Aberration and delay. The delay and aberration angles b for the light (and gravitational force) reaching the earth from the sun equal about 100 micro-radians, or some 12,000 times less than shown in the figure, so that the points S and S' both lie in the sun and are practically identical.

velocity, then by Sec. 1.3.1, the light will arrive from an aberrant angle, altering the geometrical angle (perpendicular to the velocity) by $\sin\beta$, or since β is small (3×10^{-4}), by β. The aberration is in the direction of the velocity, so that the light will arrive in the direction from S', which is parallel to, and therefore identical with, the original direction to E_1, in which the light left the sun 8 minutes ago. In other words, to first order the effect of the delay is canceled by the effect of aberration, so that the bright spot in the sky is the location of the real sun.

It is a sobering thought that when the professors are through arguing, they find the position of the sun exactly where the janitors never doubted it to be; and if we apply this janitors' principle to the fictitious celestial body Σ which travels at the same angular velocity as the earth but lies beyond it, so that its light — like the sun's gravitational force — has the same direction as sunlight, but the opposite sense, then clearly everything that has been said about the propagation of light must equally well apply to the propagation of force. In particular, the attractive gravitational force has the same aberrant direction as sunlight, that is, it arrives not exactly perpendicular to the earth's velocity, but at an angle $90°-\beta$ to it. The same argument goes for electrons orbiting round the nucleus (at zero level in the hydrogen atom, where the electron is fastest, $\beta \approx 0.007$).

This aberration of force results in a sharp difference from the conventional Newton-Coulomb Law (6), where for the case of an electron orbiting the atomic nucleus, the constant $K = 2.3 \times 10^{-28}$. This formula applies in all theories when the charges are at rest — whatever the pertinent theory understands by that. When they are moving with respect to that theory's rest standard, then in the Einstein theory, as we have just seen, (6) applies to the proton attracting the electron, but not vice versa.

Now let us look at the dynamic Newton-Coulomb Law for the force between two moving charges from the point of view of the present theory. Even for the case of a circular orbit, there will be three differences from the static case (6). First, the bunching of the electric field strength in the transversal direction (seen by the Einsteinian observer if he sits on the nucleus, but observer-independent here); second, the magnetic force between moving charges as discussed in Sec. 1.4; and third, the effect of aberration just discussed.

Let us modify (6) for these points. The bunching of the **E** lines of force for **E** in the direction of the acceleration when the latter is normal to the velocity is given by (4), so that (6) becomes

$$\mathbf{F} = \frac{K}{r^2\sqrt{1-\beta^2}}\mathbf{r}_0 \tag{7}$$

Next, we correct for the magnetic force between the two charges; according to magnetic rule no. 2, i.e., (6), Sec. 1.4, we must multiply (7) by $1-\beta^2$, resulting in

$$\mathbf{F} = \frac{K\sqrt{1-\beta^2}}{r^2}\mathbf{r}_0 \tag{8}$$

and finally, we must correct for the aberrational angle under which the nucleus acts on the electron; that angle, from the discussion above, is β, so that

$$\mathbf{F} = \frac{K\sqrt{1-\beta^2}}{r^2}(\cos\beta\,\mathbf{r}_0 + \sin\beta\,\boldsymbol{\Theta}_0) \tag{9}$$

where $\boldsymbol{\Theta}_0$ is a unit vector in the transversal direction in polar coordinates r, θ, and for the case of a circular orbit also the unit vector in the direction of the velocity.

In both applications that will be of interest, planetary systems and electron orbits, β is small: about 7×10^{-3} for the ground level in the hydrogen atom, and about 1.4×10^{-4} for Mercury, the fastest planet. For accuracy to second order, we therefore use

$$\cos\beta = \sqrt{1-\beta^2} + O(\beta^4) \tag{10}$$

and with the same accuracy, (9) simplifies to

$$\mathbf{F} = \frac{K}{r^2}[(1-\beta^2)\mathbf{r}_0 + \beta\boldsymbol{\Theta}_0] \tag{11}$$

which is an unusually radical departure from the conventional Coulomb-Newton Law (6), for a *first-order* term in β has appeared in it. For an attractive force, such as that of the sun or of the nucleus, the constant K is negative, so the θ component of this force in the second term of (11) is directed *against* the velocity. (This is not affected by a choice of coordinates since for positive K the aberration always deviates from the "true" angle in the direction of the velocity.) The first thing that must therefore be explained is why the solar system, and all of its atoms, do not collapse.

This would indeed be the case if the β^2 term were not present in the radial component. In that case the orbiting body would do work in advancing against the force (11) in the direction of its velocity, and since the system is closed, this work would have to be performed at the expense of the potential energy, i.e., by reducing the distance of the orbiting body from the attracting center. Quantitatively, an element of work performed by the force (11) over a distance

$$d\mathbf{s} = dr\,\mathbf{r}_0 + r\,d\theta\,\boldsymbol{\Theta}_0 \tag{12}$$

is

$$\mathbf{F} \cdot d\mathbf{s} = 0 \tag{13}$$

which equals zero since there cannot be any net energy change.

Substituting (11) and (12) in (13) yields an elementary differential equation with solution

$$r = r_0 e^{-\beta\theta} = r_0 e^{-\beta\omega t} \tag{14}$$

where we have assumed β constant and equal to its instantaneous value over a few turns of this inward spiral.

However, the β^2 term in (11), which has been ignored in this calculation, compensates for this effect as follows. The path (14) corresponds to an effective force pushing the orbiting body toward the center given by

$$\mathbf{F}_{in} = -m\ddot{r}\,\mathbf{r}_0 = -\beta^2\omega^2 r\,\mathbf{r}_0 \tag{15}$$

On the other hand, the β^2 term in (11) is positive (for K is negative for attraction), representing a force in the direction of $\mathbf{r}_0$, or outward. On using the formula for the angular velocity of an orbiting body (derived in textbooks of mechanics, and also in Part Two)

$$\omega^2 = \frac{|K|}{mr^3} \tag{16}$$

this term is

$$\mathbf{F}_{out} = \frac{|K|\beta^2}{r^2}\,\mathbf{r}_0 = \beta^2 m\omega^2 r\,\mathbf{r}_0 \tag{17}$$

which exactly cancels (15).

However, apart from saving the solar system and its atoms from collapse, this term is of no significance for the small values of β in planetary and atomic orbits, and we shall henceforth neglect it, leaving the modified Newton-Coulomb Law for circular orbits (and more generally for the force perpendicular to the velocity) in the form

$$\mathbf{F} = \frac{K}{r^2}[\mathbf{r}_0 + \beta\mathbf{\Theta}_0] \tag{18}$$

The radial component of the force introduces no aberration, but gives rise to delay. Let two equal charges move away or toward each other uniformly along the straight line joining them and let their instantaneous (inferrable) distance from each other be r. Then during the time the force has propagated over the distance r, namely the time $t = r/c$, the distance between the charges will have increased by $\dot{r}t = \dot{r}r/c$. Therefore the force propagating from the source charge will act as from the point when it was emitted, not from the point where the object charge is at the time of arrival. That is equivalent to modifying the distance between the charges by a factor $(\dot{r}/c)$. Thus the full version of the modified Newton-Coulomb Law (comparable to the Liénard-Wiechert formula) to first order in β is

$$\mathbf{F} = \frac{K}{r^2(1 - \dot{r}/c)^2}[\mathbf{r}_0(1 - \beta^2) + \beta\mathbf{\Theta}_0] \tag{19}$$

However, we shall need this case only once, namely in the advance of Mercury's perihelion in Part Three. Otherwise the modified Newton-Coulomb Law we shall use throughout the book will be in the form (18).

The original Newton-Coulomb Law (6) will be seen identical with (11), (18) or (19) for $c = \infty$, corresponding to Instant Action At a Distance (IAAD).

To summarize, the present theory assumes that forces propagate with velocity c from their sources, that Newton's Laws and the Maxwell equations are valid when all velocities are referred to the local force field rather than to an observer, and that the relativity principle is valid in Euclidean space and unreformed time. This leads formally to the same expressions for mass, momentum and energy, and to the same relations among these three as in the Einstein theory, but the corresponding effects are rooted in the phenomena themselves, independently of any observer's location or perceptions.

It will now be shown that the theory explained so far will explain all observed effects invoked as proofs of the Einstein Theory. The two additional effects that have hitherto remained unexplained will be discussed in Parts Two and Three.

1.9 The Electromagnetic Evidence

To show that the proposed theory does not contradict the experimental evidence in the field of electromagnetism, it must be shown that

1) the experiments confirming the Einsteinian formulas for mass, energy and momentum are consistent with the proposed theory (in which these formulas are, as explained in Sec. 1.6, only *formally* identical);

2) that the evidence purporting to demonstrate time dilation has been misinterpreted;

3) that the electromagnetic equations of moving material media remain valid in the present theory; and

4) that the Maxwell equations and the Lorentz force are invariant to the *Galileian* transformation once all velocities in them have been referred to the field rather than to the observer.

As for length contraction, there is no direct experimental evidence for it, and therefore no need to refute or reinterpret it. The indirect evidence comes from experiments such as that by Michelson and Morley; but as we have seen, this experiment is consistent with at least four different theories, of which the proposed theory is one.

1.9.1. Mass, Momentum and Energy

The crucial relations of Einsteinian dynamics that have been confirmed by experiment are those involving mass, energy and their relation to momentum, given by (7), (11) and (15), Sec. 1.7, respectively. However, in the Einsteinian interpretation, $\mathbf{v}$ is understood to mean the velocity with respect to the observer rather than with respect to the local force field.

All that needs to be shown, therefore, is that in all of these experiments the observer was at rest with respect to the local force field, so that the experiments cannot decide whether the effective (effect-producing) velocity is that with respect to the observer or that with respect to the field.

For example, one of the ways of measuring mass at high velocities is to let a charged particle traverse a magnetic field at right angles. The Lorentz force $q\mathbf{v}\times\mathbf{B}$ will curve the path of the particle, which will balance this force with its inertial reaction (centrifugal force) mv^2/r. From the equality, the radius of curvature is

$$r = \frac{m_0 v}{qB\sqrt{1-\beta^2}} \tag{1}$$

and all quantities in this relation can be measured. The relation has been confirmed with protons for β as high as 0.81 [Zrelov *et al.* 1958].

In all of these experiments using a magnetic field, the latter is, of course, produced by wire-bound currents (permanent magnets would also be electrically neutral). As explained in Sec. 1.4, for such magnetic fields the local force field is that of the electrically neutral conductor, that is, the gravitational field in which the laboratory is also at rest. These experiments are therefore consistent with both theories and unable to test the difference between them.

The dependence of kinetic energy on velocity

$$E_{kin} = mc^2 \left(\frac{1}{\sqrt{1-\beta^2}} - 1 \right) \tag{2}$$

derived in (11), Sec. 1.7, has been confirmed, for example, by measuring the heat dissipated in the water tank in which the high-velocity electron beams of linear accelerators are dumped. A recent experiment by Waltz *et al.* [1984] is impressive, not so much for its accuracy (an error of 30%) as for its high value of β (0.9995) and of the dissipated beam power (up to 3.5 kW).

In a linear accelerator, particles are accelerated by a series of "gaps" across which the accelerating voltage is supplied by a radio-frequency traveling wave arriving at successive gaps simultaneously with (or just slightly ahead of) a bunch of particles. It might therefore be thought that the "velocity with respect to the local force field" should be the velocity of the particles with respect to the traveling wave, which would be close to zero.

Not so: here and in all other cases, velocity with respect to the local field means the velocity of the particle with respect to the lattice of equipotential/line-of-force intersections in its immediate neighborhood, without regard to how this field got there. The particles are accelerated in steps as they traverse the gaps, with no significant electromagnetic field or acceleration in their flight from gap to gap. During the short time that they traverse a gap, the electric field accelerating them is produced by the charges on the opposite ends of the *stationary* gap, not by the fields in other places, which are as irrelevant as their mutual relationship that constitutes the traveling wave. The field within the stationary gap is nailed to it as securely as if it were produced by a battery switched on and off at the proper times. Thus the "local force field" is stationary in the laboratory frame, which is the rest frame for the observer. Once again, this type of experiment is consistent with both theories and unable to test the difference between them.

1.9.2. Champion's Experiment

It should be noted that a test of a relation like (1) of the preceding section, though consistent with both theories, is not the most convincing thing in the world. The velocity is not directly measured, but inferred from the electromagnetics that the test is to confirm. In addition, the test *assumes* the conservation of charge, for it is a matter of interpretation, not a matter of measurement, whether the square root divides the rest mass or multiplies the "rest charge," which in some other theory (not Einstein's or mine) might not be constant. This is an objection which applies to all experiments involving the mass-to-charge ratio, and this includes a large number, perhaps even a majority, of experiments claiming to prove the velocity dependence of mass. In reality they prove nothing but the velocity dependence of the mass-to-charge ratio.

A clean (or at least, cleaner) experiment would demonstrate the mass-energy-momentum relations independently of the value of charge or velocity used. There is such an experiment, an effect apparently first noted by Champion [1932], for in its simplest form it measures nothing but the change of an angle — the angle of the paths of two electrons after collision. In IAAD (instant-action-at-a-distance) mechanics, the tracks of these electrons, which can be recorded in a cloud chamber, should be perpendicular, but at high velocities they were observed to conclude an acute angle. By 1935, velocities corresponding to $\beta = 0.968$ had been achieved [Tonnelat 1959], and the good agreement with theory was widely interpreted as a confirmation of the Einstein theory.

The experiment confirms the relations derived in Sec. 1.7 when interpreted by the present theory. However, it is my belief that the Einstein theory comes through this test with less than flying colors, as discussed in the following.

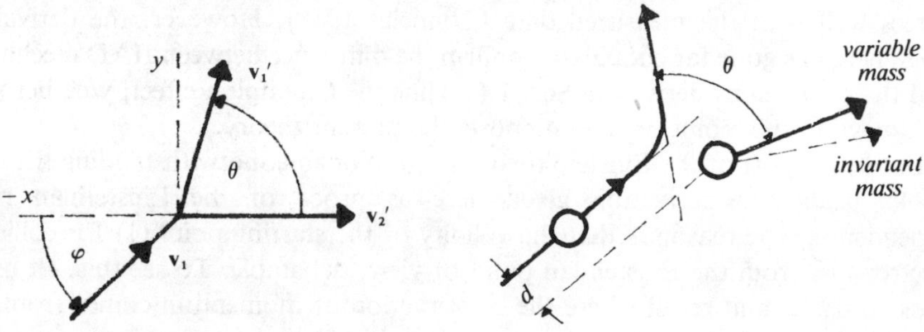

(a) symbols and geometry for (1) (b) repulsion at a distance

Champion's experiment

Let two electrons (or billiard balls, for that matter) collide; let one of them be originally at rest at the origin, hit by the other with momentum **p**. We orient the x-axis along the path of one of the balls after collision and denote the momenta after collision with primes. Then from the conservation of momenta along the x and y axes we have

$$\left.\begin{aligned} p_1 \cos\varphi &= p'_1 + p'_2 \cos\theta \\ p_1 \sin\varphi &= p'_1 \sin\theta \end{aligned}\right\} \tag{1}$$

Squaring and adding yields

$$p_1^2 = p'^2_1 + p'^2_2 + 2p'_1 p'_2 \cos\theta \tag{2}$$

Using (15), Sec. 1.7, in the form

$$\frac{p^2}{c^2} = m^2 - m_0^2 \tag{3}$$

and from the conservation of energy (dividing by c^2)

$$m_1 + m_2 = m'_1 + m'_2 \tag{4}$$

we find after some algebra

$$p'_1 p'_2 \cos\theta = c^2(m'_2 - m_0)(m'_1 - m_0) \tag{5}$$

In IAAD mechanics, the masses on the right are all equal, so that the right side vanishes, whence $\cos\theta = 0$ and $\theta = \pi/2$. But when mass is a function of velocity as in (5), Sec. 1.6, both parentheses on the right are positive, as are the momenta on the left; hence $\cos\theta$ is positive and θ is acute (less than $\pi/2$).

Some more algebra will actually express $\tan(\varphi - \theta)$ as a function of β, and this agrees well with the measured data [Tonnelat 1958]. However, the derivation given here has gone far enough to confirm the difference between IIAD mechanics and the expressions derived in Sec. 1.6. Thus the Champion effect, whether pursued beyond this point or not, supports the present theory.

But its support of the Einstein theory is questionable, notwithstanding the textbooks using the derivation given here as proof of the Einsteinian mass dependence. The reason is that the validity of the starting point (1) for colliding electrons is, from the Einsteinian point of view, debatable. To see this, let us go back to basics and recall where the conservation of momentum comes from.

Let a system of bodies (such as *charged* billiard balls) be subject to external forces (e.g., currents flowing nearby and friction on the billiard table). Let the external forces on the kth body be $\mathbf{F}_k$, and let the internal forces by which any two bodies act on each other be $\mathbf{F}_{kj}$ and $\mathbf{F}_{jk}$. Then since no body acts on itself, all $\mathbf{F}_{jj} = 0$, and integrating over time from t_1 to t_2, we have

$$\int_{t_1}^{t_2} \left(\sum_k \mathbf{F}_k + \sum_j \sum_k \mathbf{F}_{jk} \right) dt = \sum_k [m_k \mathbf{v}_k(t_2) - m_k \mathbf{v}_k(t_2)] \tag{6}$$

provided the double sum on the left vanishes due to action and reaction being equal and opposite, i.e., provided that $\mathbf{F}_{jk} = \mathbf{F}_{kj}$. If there are no external forces, then the left side vanishes completely, and since the times t_1, t_2 are arbitrary, the remainder states the conservation of momentum at all times.

But the words printed in italics, which are always valid in the present theory, do not necessarily hold in the Einstein theory. They hold for the collision of uncharged billiard balls, when momentum is transferred by actual contact at a point where simultaneity holds for all observers. They do not hold for an electron, because it does not wait in space, nailed to its coordinates while Coulomb's Law is suspended, until it is bodily hit by another electron. What happens is that an electron is repelled by the approaching electron at a distance; the two come to within a minimum distance, but not into contact, and they continue to repel each other as they recede from each other. During the entire process, which is studied in plasma physics and particle scattering, the two interact *at a distance*. But as we have seen in (4) and (5) of Sec. 1.8, the Einstein theory does *not* recognize the equality of action and reaction at a distance: the force exerted by a particle at rest on a moving particle is not at all the same as the force exerted by the moving particle on the one at rest.

I have no doubt that the Einstein theory can explain things as it always does. Perhaps the asymptotes of the curved trajectories are as good as straight lines from an equivalent collision; perhaps the whole thing can be conjured away in the opaque acrobatics of four-vectors and world lines.

But the fact remains that the Einstein theory has some explaining to do; for a theory that does not recognize the equality of action and reaction cannot, without apology, invoke the conservation of momentum.

1.9.3. Time Dilation: Ives-Stilwell, Mesons and Clocks Around the Globe

According to Einstein, a clock ticks more slowly for an observer who passes it with velocity $\mathbf{v}$ than for one who is at rest with respect to it. (Both observers compare its reading to their identically constructed electronic wristwatches, say).

We are offered three types of experimental proof for this phenomenon: the Einsteinian Doppler effect, the rate of decay of fast moving mesons, and the transport of an accurate clock round the globe.

What all three techniques have in common is the failure to ask, let alone answer, the crucial question: is the measured effect something that is dependent on the observer, or is it something that changes the clock?

To see the difference, imagine two identical *pendulum* clocks whose time readings are compared after one of them has been transported round the globe.

If flown eastward, the transported clock's reading will be fast; flown westward, it will be slow.

Time dilation?

No: the period of a pendulum varies inversely as the square root of the downward force on it, and that force is the vector sum of gravitational attraction and the centrifugal force due to the earth's rotation. As pointed out by Barnes [1983], the centrifugal force must necessarily, if only very slightly, increase when the clock is moved eastward, because its angular velocity about the earth's center increases; and it must decrease when transported westward, against the earth's rotation. This is an *inherent* change, one that an observer traveling with the clock (i.e. at rest with respect to it) could measure by comparison with an equally accurate wrist watch – if it is unaffected by centrifugal force.

I do not, of course, propose this as an explanation of the alleged time dilations; I mention it as an illustration of an inherent change in a clock which might easily be mistaken for a change in the flow of time.

In this and all other cases we must first check by a control experiment whether the rate of the clock has changed inherently, as measured by a co-traveling observer at rest with respect to the clock, before we check for any Einsteinian observer-dependent effects. *In none of the three techniques has this been done.*

Such a control experiment performed by observers (measuring instruments) traveling with hydrogen ions or mesons as they traverse a gravitational field at a significant fraction of the velocity of light are beyond contemporary feasibility; so it is tacitly *assumed*, without the slightest proof, that there are no such inherent changes, and all observed changes are ascribed to the observer's velocity.

Consider the Ives-Stilwell [1938, 1941] experiment on the Doppler effect of a fast moving source (light-emitting hydrogen ions in canal rays), which is concep-

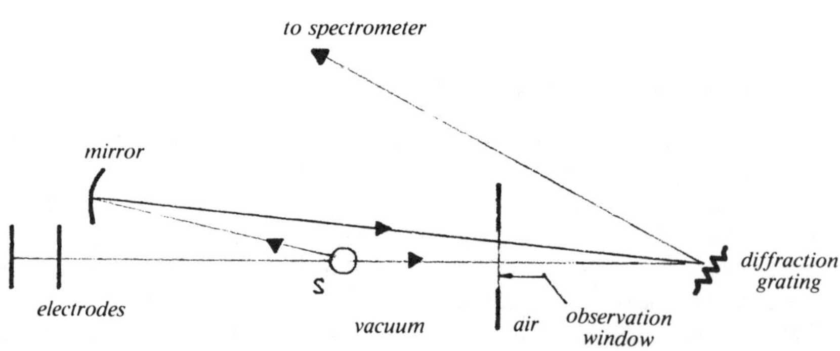

The Ives-Stilwell experiment [1938, 1941]. The ions S are generated to the left of the figure, accelerated by the electrodes, and pass through a hole in them to the space on the right. Their light reaches the diffraction grating from an approaching source through the observation window directly, and from a receding source via the concave mirror M, whose axis is only 7° off the velocity direction. The grating is the dispersive element of the spectroscope, whose telescope and photoplates are not shown.

tually the simplest of the three types; it is also very impressive because its result depends only on a comparison of spectroscope readings, not on inferred velocities.

As shown in the figure, the grating of the spectroscope is reached by the light emitted by fast moving hydrogen ions directly in the forward direction, and via the mirror in a direction making an angle of only 7° with the velocity. Thus the spectroscope measures the Doppler-shifted wavelengths of the radiation emitted by an approaching and a receding, yet identical, source.

The classical Doppler effect for a source moving with velocity $\mathbf{v}$ in a medium in which the observer is stationary (also applicable to the propagation of light through a gravitational field) is found by elementary trigonometry. The time difference between two successively received wave crests emitted with period T_0 is

$$T = T_0 + \frac{R}{c} - \frac{r}{c} \tag{1}$$

where R and r are the distances of the source from the observer at the moments when the crests were emitted. For $\lambda << r$, the emitted and received wavelengths λ_0 and λ are therefore related by

$$\lambda = \lambda_0(1 - \beta\cos\theta) \tag{2}$$

where $\beta = v/c$, and θ is the angle made by $\mathbf{v}$ and the direction of propagation to the receiver.

The Einsteinian Doppler effect, on the other hand, leads to

$$\lambda = \frac{\lambda_0(1 - \beta\cos\theta)}{\sqrt{1-\beta^2}} \approx \lambda_0(1 - \beta\cos\theta + \tfrac{1}{2}\beta^2) \tag{3}$$

Let us now write (2) and (3) for small as

$$\lambda = \lambda_0(1 - \beta\cos\theta + k\beta^2) \tag{4}$$

where $k = 0$ in the classical, and $k = ½$ in the Einstein theory. The Ives-Stillwell experiment is based on the asymmetry of (4): when the sign of β is changed (i.e. the backward ray is considered instead of the forward ray), and the two resulting wavelengths averaged, the terms in β will cancel, but the ones in β^2 will remain. The Doppler-shifted wavelength is

$$\Delta\lambda = \frac{\lambda - \lambda_0}{\lambda_0} = -\beta\cos\theta + k\beta^2 \tag{5}$$

and this measured by the spectroscope. The two Doppler-shifted lines, one from the approaching and one from the receding ray, correspond to $+\beta$ and $-\beta$ and are displaced to either side of the "rest" spectral line. When the two shifts are averaged, we then have from (5)

$$\Delta_2\lambda = \tfrac{1}{2}[\Delta\lambda_1 + \Delta\lambda_2] = k\beta^2 \tag{6}$$

We can now combine (5) and (6) into the relation

$$\Delta_2\lambda = k(\Delta\lambda)^2 \tag{7}$$

in which the the bone of contention k can be checked by measuring only wavelengths, unaffected by (reasonable) errors in voltage, velocity, or other error-prone quantities: the spectroscope measures the two shifted lines as in (5), the average as in (7) is then examined on the photographs under a microscope, and the results are plotted for comparison with (7).

Working on the borders of the then feasible technology (mechanical micrometer, all-day exposures), Ives and Stilwell established that $k = ½$. This was confirmed by Mandelberg and Witten [1962] with the technological advance of two decades, and as this book was readied for press, news came that MacArthur *et al.* [1986] had obtained further confirmation at $\beta = 0.84$.

To Einsteinians this is proof of time dilation; to me it is proof that particles traversing a gravitational field radiate, in their own rest frame, an inherent frequency lowered by $½\beta^2$ — reminiscent of the seeming time dilation of the pendulum clock. This is the frequency an observer sitting on a moving particle would measure, and this is then Doppler-shifted by classical rules to yield the result measured by Ives and Stilwell.

Without a check of what frequency is radiated inherently in the source's own frame, the Ives-Stillwell result remains ambiguous, and would not have to be accepted even if there were no alternative explanation.

But there is one.

When a hydrogen atom moves through a gravitational field, then by our basic assumption, force propagates with velocity **c** with respect to the gravitational field; thus the Coulomb force between nucleus and electron will be subject to delays and aberrations. The delay, if any, has no effect, since the radius of the electron orbit remains unchanged. In calculating the aberration, we ignore the aberration due to the electron's orbital velocity — not because it is negligible, but because we are looking for the *additional* aberration that sets in when the atom moves as a whole. From elementary geometry and Sec. 1.3.1 we find the aberration angle

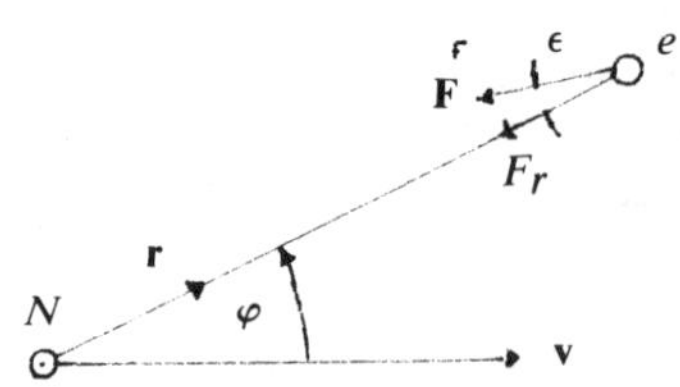

Aberration and frequency shift

$$\sin\epsilon = \beta\cos\varphi \tag{8}$$

where φ defines the electron's instantaneous position on its orbit. The force in the radial direction (toward the nucleus) is therefore reduced by

$$\cos \epsilon = \sqrt{1 - \beta^2 \cos^2 \varphi} \approx 1 - \beta^2/4 \tag{9}$$

where $\cos^2 \varphi$ has been averaged over the orbit. This has the same effect as if the square of the charge were reduced by that amount; we can therefore calculate force, energy or other quantities by using the fictitious charge

$$q = q_0(1 - \beta^2/8) \tag{10}$$

where q_0 is the equivalent "rest charge." The radiated frequency (or energy quantum $h\nu$) is proportional to the fourth power of this charge [as derived in (18), Sec. 2.5, or found in any physics handbook], so that the *inherently radiated* frequency is

$$\nu = \nu_0(1 - \tfrac{1}{2}\beta^2) \tag{11}$$

and this is then shifted by the classical Doppler effect to yield exactly what was measured by Ives and Stilwell, or when inverted, to give the time difference on a transported cesium clock.

For radioactive decay [Frisch and Smith 1963], the timing mechanism is unknown, but here again energy is proportional to the average frequency of disintegrations; the assumption that (11) remains valid for this case is no more arbitrary than assuming (as is done in orthodox physics) that radiated frequencies or radioactive decay remain inherently unchanged when the corresponding atoms traverse a force field with a velocity approaching that of light.

Thus we again obtain the same results as in the Einstein theory.

To summarize: the experimental evidence on alleged time dilation overlooks the crucial issue: is it time or the clock that is affected? It is a special case of a more fundamental question: should physics seek to understand objective reality or should it describe an observer's perceptions?

1.10. Galileian Electrodynamics

1.10.1. The Maxwell Equations and the Lorentz Force

There is something puzzling about the Maxwell equations: they grew out of Faraday's concept of lines of force repelling each other as they weave their way through the ether; this concept has been totally abandoned, yet the Maxwell equations have remained valid through all types of ether — elastic, rigid, stationary, entrained, partially entrained — and they remain valid in the Einstein theory. It is not the usual fate of a flower to bloom on when the soil is removed or shown never to have existed.

The answer to the puzzle lies in what the computer sages call "transportability:" a good computer language, for example, will run under various operating systems when it has comparatively few routines that must be adapted to the system and is otherwise self-contained.

The Maxwell equations give the relations among four vectors, **E**, **D**, **B**, **H**. These vectors cannot manifest themselves until they act on charged matter; if the Maxwell equations were not tied to charge and matter, they would be a meaningless abstraction. But the bridge to charge and matter is a narrow one (to be discussed in a moment), and the bridge from charge to force is given by a single equation, namely the Lorentz force

$$\mathbf{F} = q(\mathbf{E} + \mathbf{v} \times \mathbf{B}) \tag{1}$$

in which the magnetic induction **B** is not an independent vector, but is (for a moving point charge) derived from the electric field by

$$\mathbf{B} = \frac{\mathbf{v} \times \mathbf{E}}{c^2} \tag{2}$$

Here, very typically, **v** is the velocity of the moving charge q with respect to an unspecified rest standard: the stationary ether, the entrained ether, the observer, the field of the other charge(s) — the Maxwell equations care no more than a transportable computer program cares what operating system it is running under.

Thus, the Maxwell equations proper are a self-contained floating island that can be linked to the mainland of charge and matter by various bridges involving various velocities, and this solves the puzzle why the island survived when the various mainlands went under.

Let us now look for the bridges that tie the floating abstractions to the firm mainland of charge and force. We write the Maxwell equations (for piecewise homogenous, isotropic, non-pathological media) as

$$\nabla \times \mathbf{E} + \frac{\partial \mathbf{B}}{\partial t} = 0 \tag{3}$$

$$\nabla \times \mathbf{B} - \frac{1}{c^2}\frac{\partial \mathbf{E}}{\partial t} = \mu \mathbf{J} \tag{4}$$

$$\nabla \cdot \mathbf{E} = \frac{\rho}{\epsilon_0} \tag{5}$$

$$\nabla \cdot \mathbf{B} = 0 \tag{6}$$

These four equations contain only two quantities that inconspicuously provide a bridge to charge and matter, namely the charge density ϱ and the current density **J**. If we set these two to zero, we are in the well known textbook case of "a space without charges or currents," the obvious implication being that we are in a space in the *neighborhood* of charges and currents.

Note two points about these two quantities, without whose presence (immediate or distant) Maxwell's equations become a meaningless torso:

First, both are velocity-dependent (the velocity-dependence of ϱ is attributed to that of volume in the Einstein theory, and is caused by a charge redistribution in mine). They are are tied together, not by a natural law, but by a *definition*, the definition of current density:

$$\mathbf{J} = \rho \mathbf{v} \tag{7}$$

making the bridge even narrower. (A more general definition, expressing the invariance of charge, is possible, but will not be needed here.)

Second, in the Einstein theory, these two quantities are *not* invariant to the Lorentz transformation (even though the equations involving them are): charge density is charge per volume, where charge is an invariant, but volume has one dimension that shrinks with velocity. Current density is modified even more drastically in the Einstein theory, for velocity is transformed in a more complicated way than length. Hence the bridge (7) is no longer a simple relation in the Einstein theory, but a mask hiding an ugly complication.

There are three more quasi-bridges to charge and matter, namely the constituent equations

$$\mathbf{D} = \epsilon \mathbf{E} \tag{8}$$

$$\mathbf{H} = \frac{\mathbf{B}}{\mu} \tag{9}$$

to which one may add Ohm's Law, if applicable,

$$\mathbf{J} = \sigma \mathbf{E} \tag{10}$$

and all of this, of course, remains an abstraction until it has manifested itself by the force (1).

The reason why I have called the constituent equations *quasi*-bridges from fields to charged or uncharged matter is that for moving media they are not really bridges; they are no more than formulations of the problem, as evident from the following.

Maxwell's equations (3) through (6) are straightforward only for fields in unlimited free space; the behavior of fields in matter, and especially in moving matter, has been pushed off into the constituent equations (8) through (10), which give the relation between the fields inside matter (**D**, **H**) and the fields producing them, quite often in the free space outside it, (**E**, **B**). As long as this matter is at rest, then we have nothing to talk about, for the "rest" permittivity ϵ and the "rest" permeability μ are quite uncontroversial, and we need neither Lorentz nor Galileo for electrodynamics where the flow of charges as electric currents is the only thing that moves. (Let us not not complicate matters by the conductivity σ, which can, in the most important case of harmonic time variation, simply be absorbed into the imaginary part of a complex permittivity). But when we have a moving medium, what happens to the permittivity and permeability – what happens to the fields inside matter – when we are at rest and the medium is moving?

To this one and only question of importance for moving media, the constituent equations answer with two shoulder-shrugging Greek symbols, ϵ and μ, implying a relation between the fields in the moving medium and the fields outside producing them, but telling us nothing about what permittivity and permeability "on the move" stand for, much less how to find them.

The relation between the field **D** inside a medium and the field **E** outside it is not at all simple when the medium is moving, nor is it simple between **H** and **B**; in fact, it is best to abandon the concepts of ϵ and μ in (8) and (9), which become tensors, for **D** and **E** (just as **B** and **H**) inside a moving medium no longer have the same direction even if the medium is isotropic.

The Einstein theory supposedly first fully solved the problem of electromagnetic fields in moving media; and we shall examine this claim next.

1.10.2. Electromagnetics of Moving Media

In his classic paper [1905a], Einstein showed that the Maxwell equations are Lorentz-invariant when applied to individual moving charges (with the effect-producing velocities assumed to be those with respect to an observer). But it fell to his disciple Hermann Minkowski (1864-1909) to apply Einstein's theory to moving matter, that is, to find the relation between the "driving" fields **E**, **B** and the "driven" fields inside matter, **D**, **H**. For moving matter, such a relation is implied, but not explicitly given, by the constituent equations (8) through (10) of the preceding section. Minkowski, the mathematician who introduced space-time ("Henceforth space by itself, and time by itself, are doomed to fade away into mere shadows. . ."), found the solution in 1908 via six-vectors and their space-time components; his derivation is given in [Sommerfeld 1964] or [Penfield and Haus, 1967]. One of the simpler ways of writing the required relations is

$$D_{\|} = \epsilon E_{\|} \tag{1}$$

$$B_{\|} = \mu H_{\|} \tag{2}$$

$$\left[1 - \frac{\epsilon\mu}{\epsilon_0\mu_0}\beta^2\right] D_{\perp} = \epsilon(1-\beta^2)E_{\perp} + (\epsilon\mu - \epsilon_0\mu_0)\,\mathbf{v}\times\mathbf{H} \tag{3}$$

$$\left[1 - \frac{\epsilon\mu}{\epsilon_0\mu_0}\beta^2\right] B_{\perp} = \mu(1-\beta^2)H_{\perp} - (\epsilon\mu - \epsilon_0\mu_0)\,\mathbf{v}\times\mathbf{E} \tag{4}$$

$$\left.\begin{aligned} (\mathbf{J}-\rho\mathbf{v})_{\|} &= \sigma\sqrt{1-\beta^2}(\mathbf{E}+\mathbf{v}\times\mathbf{B})_{\|} \\ (\mathbf{J}-\rho\mathbf{v})_{\perp} &= \frac{\sigma(\mathbf{E}+\mathbf{v}\times\mathbf{B})_{\perp}}{\sqrt{1-\beta^2}} \end{aligned}\right\} \tag{5}$$

where $\|$ and $\perp$ denote the components parallel and perpendicular to the velocity.

However, to my knowledge these equations have never been verified to second order, nor is it easy to do so: they cannot, like the mass dependence, be tested by particle accelerators, for they involve the fields in macroscopic, ponderable matter. The electrodynamics of moving media based on (1) through (5) (or one of several other formulations) is an esoteric, highly theoretical field, which is not without problems (such as the equivalence of various formulations, see [Penfield and Haus, 1967]) and which gives little physical insight.

On the other hand, for slowly moving media (first-order β), the corresponding equations can be derived without the Einstein theory, are verified, and do provide physical insight, for they rest on simple principles. For example, the polarization **P** in a moving medium induced by an external stationary field **E** will result in charges that are bound within the medium, yet moving with respect to the rest frame, having the effect of a current. Such a physically founded derivation is no longer easy to find, but it does exist: for non-magnetic media, see [Becker 1964], while for the general case I know only of a Russian source [Tamm 1954].

These derivations without use of the Lorentz transformation lead to the same result as (1) through (5) with $\beta^2=0$, as they must, since the Galileian and Lorentz transformations merge under that condition. Thus for slowly moving media we have from (3) to (5)

$$\mathbf{D} = \epsilon\mathbf{E} + \frac{n^2 - 1}{c^2}\mathbf{v} \times \mathbf{H} \tag{6}$$

$$\mathbf{B} = \mu\mathbf{H} - \frac{n^2 - 1}{c^2}\mathbf{v} \times \mathbf{E} \tag{7}$$

$$\mathbf{J} - \rho\mathbf{v} = \sigma(\mathbf{E} + \mathbf{v} \times \mathbf{H}) \tag{8}$$

Since these are the only experimentally verified equations for moving media, there is no need to attempt the derivation of (1) through (5).

However, we will pursue another point that will turn out to be instructive and relevant to our story: let us see whether we can simplify (6) through (8) further by looking for equivalent electromagnetic parameters ϵ, μ, σ, that would express these equations in the more familiar form of an equivalent *stationary* medium

$$\mathbf{D} = \epsilon'\mathbf{E} \tag{9}$$

$$\mathbf{B} = \mu'\mathbf{H} \tag{10}$$

$$\mathbf{J} = \sigma'\mathbf{E} \tag{11}$$

where, in (8) and (11), we have assumed the absence of free charges ($\varrho=0$).

The problem was posed and investigated some years ago in a brief paper [Beckmann 1970]. The conditions under which such equivalent parameters of a stationary medium exist were then found to be

$$\beta n \ll 1, \qquad \mathbf{E} \cdot \mathbf{H} = 0 \tag{12}$$

in which case the only $\mathbf{v}$ component that need be considered is

$$\mathbf{v} = v \cos\theta\, \mathbf{s}_0 \tag{13}$$

where $\mathbf{s}$ is a unit vector in the direction of the Poynting vector $\mathbf{E}\times\mathbf{H}$ and θ is the angle made by that vector and the velocity $\mathbf{v}$.

The second condition in (12), $\mathbf{E}$ perpendicular to $\mathbf{H}$, is fulfilled in the most important application of this type of problem, namely the propagation of electromagnetic waves, and some other problems as well. If the fictitious, stationary medium is to be equivalent to the real, moving medium, the two media must obviously have the same impedance

$$\sqrt{\frac{\mu'}{\epsilon'}} = \sqrt{\frac{\mu}{\epsilon}} \tag{14}$$

From (12), (13), (14), we have

$$\mathbf{v} \times \mathbf{E} = c\beta \cos\theta \sqrt{\frac{\mu}{\epsilon}}\, \mathbf{H} \tag{15}$$

$$\mathbf{v} \times \mathbf{H} = -c\beta \cos\theta \sqrt{\frac{\epsilon}{\mu}}\, \mathbf{E} \tag{16}$$

and when these two are substituted in (6) through (8) using $n = c\sqrt{(\epsilon \upsilon)}$, we obtain by comparison with (9) through (11) the required equivalent parameters

$$\epsilon' = \epsilon \left(1 - \frac{n^2 - 1}{n} \beta \cos\theta\right) \tag{17}$$

$$\mu' = \mu \left(1 - \frac{n^2 - 1}{n} \beta \cos\theta\right) \tag{18}$$

$$\sigma' = \sigma(1 - \beta n \cos\theta) \tag{19}$$

which solve the problem.

Now let us find the velocity of propagation of electromagnetic waves in this new, equivalent medium.

Using the first of the conditions (12), we can write (17) and (18) as

$$\epsilon' = \frac{\epsilon}{1 + \kappa\beta n \cos\theta} \tag{20}$$

$$\mu' = \frac{\mu}{1 + \kappa\beta n \cos\theta} \tag{21}$$

where

$$\kappa = 1 - \frac{1}{n^2} \tag{22}$$

Hence

$$n' = \frac{n}{1 + \kappa\beta n \cos\theta} \tag{23}$$

and the velocity of propagation is

$$u' = cn' = \frac{c}{n} + \kappa v \cos\theta \tag{24}$$

This is the velocity of propagation of the electromagnetic wave through a medium which moves with velocity υ with respect to the reference frame, such as that of the laboratory observing the light propagating through the flowing water of Fizeau's 1851 experiment; for (22) is nothing but Fresnel's coefficient of drag.

The Einstein theory derives this crucial coefficient as the first term in the series expansion of the velocity-addition theorem based on the Lorentz transformation, but as explained in Sec. 1.3.2, Hoek derived it in 1868 from the null effect in his experiment on purely optical grounds without the need for an ether.

We now get the "etherless" Fresnel coefficient a second time from strictly electromagnetic considerations. There is no experimental evidence to decide whether the velocity $\mathbf{v}$ means the velocity of matter with respect to the local field, as we assume, or whether it should be referred to the observer, as in the Einstein theory, for clearly in all Fizeau-like experiments the observer was at rest with respect to the field (for the case of light, the gravitational field).

Once again the two theories lead to the same result, but once again, the electromagnetic derivation is based on tangible fields and impedances, whereas Einstein's derivation (which, incidentally, runs into some difficulties for non-zero angles between the velocity of the medium and that of the propagating wave) is based on an abstract addition of velocities in redefined space and time.

1.10.3. Invariance of Relative Velocities

We must be careful about what is meant by the statement that the Maxwell equations are invariant with respect to the Lorentz transformation. It means that if we transform the Maxwell equations from one system of space-time coordinates to another (each serving as the rest frame for an observer), we obtain equations of the same form in the new coordinate system when we use the Lorentz transformation.

That statement is perfectly true; but if the effect-producing velocities are not those referred to an observer, then it is also trivial and irrelevant.

It is trivial because the Lorentz transformation is founded in equations such as (7) and (9), Sec. 1.6: if the Lorentz transformation was made to fit the Maxwell equations, it is not surprising that the Maxwell equations fit the Lorentz transformation.

More important, it is irrelevant. The only velocities inherent in the Maxwell equations and the Lorentz force are those associated with current density ($\varrho\mathbf{v}$), with the magnetic field ($\mathbf{v}\times\nabla\phi/c^2$), and with the magnetic force ($q\mathbf{v}\times\mathbf{B}$). If these velocities are responsible for the pertinent phenomena when they are referred to to the local field rather than to an observer, then in any observer's coordinates they will remain invariant as *the simple vector difference between charge velocity and field velocity* in those coordinates, and the transformation that preserves a simple difference in velocities is the Galileian transformation. The same ultimate forces must therefore act in all moving observer's frames; what pretty patterns of equations for field vectors (other than forces) are preserved by what other pretty transformation does not matter.

We must, of course, distinguish between velocities that affect phenomena in which the observer is not involved – such as potential energy or hydraulic friction – and those that modify the transmission of information, force or energy to a moving observer. The latter category includes aberration and the Doppler effect in all its forms.

For example, if I walk toward a charged capacitor, I see its charges as currents flowing in my egocentric coordinates, and Einstein interprets this as part of the electric field changing into a magnetic field about these currents. But in the present theory I see not only a velocity of the charges (i.e., currents), but also a velocity of the electric field between the capacitor's plates, and since in the present theory the effect-producing, determining, applicable velocity is that of the charge on one plate *in the field of the other*, the Galileian transformation will make the effect-producing velocity equal to the difference between the two, which remains zero as before. This strengthens my conviction that the force between the plates of a capacitor, measured in unreformed space and time, cares very little about observers observing that force, even if they travel past the capacitor in a spaceship at half the velocity of light.

As for electromagnetic Doppler effects and aberrations, which do involve the observer's motion, none of the available evidence – such as the Ives-Stilwell experiment – contradicts the assumption that the effect-producing velocity is again that with respect to the local force field, which (with present technology) means the gravitational field.

We have here treated the Doppler effect and aberration as phenomena in which the observer's motion is directly involved as part of the phenomenon. On the other hand, there is no *a priori* reason why the force mutually attracting two electric charges should have the slightest dependence on an observer observing it. Most people who have not studied the Einstein theory in detail would probably "instinctively" agree with the last two statements, of which the second contradicts the Einstein theory. It is, in fact, due to such contradictions that many people "instinctively" distrust the Einstein theory in spite of eighty years of its continued successes.

But instinct is not what science is made of, and it is not easy to find a non-tautological and unambiguous criterion of what distinguishes the Doppler effect from the force between two charges as far as dependence of the observer's velocity is concerned. But at *low* velocities (with no more than first order β significant) the experimental evidence, if nothing else, shows a very striking difference between the two: the observer's velocity is what the Doppler effect is made of, but it is irrelevant to the force between two charges. And while I admit that this difference need not necessarily hold for high velocities also, I do maintain that if it does not, then the burden of proof – or at least the reponsibility to defend itself against rival theories – is on the theory which imposes such a difference on us.

Now let us turn to those phenomena for which there is no reason to believe that they are in any way physically connected to the velocity of an observer, although of course, that velocity will occur in the coordinates which he chooses to describe, but not to change, the phenomenon he is observing. As a criterion of classification, I will use the laws governing the phenomenon at low velocities: among those not affected by the observer's velocity are the force between two charges, hydraulic friction, the length of all time intervals, and many others. Among those that *are* velocity-dependent by this criterion are the Doppler effect, aberration, momentum, kinetic energy, and *not* many others.

By this criterion, if by nothing else, the forces and fields described by the Maxwell equations belong to the first group — especially when it is realized that there is no experimental evidence to differentiate between velocities with respect to an observer and those with respect to the local field, such as the gravitational field for magnetic force between two currents.

Seen through the eyes of the proposed theory, therefore, the invariance of the Maxwell equations to the Lorentz transformation *with all velocities referred to the observer* is undisputed, but immaterial, because that invariance involves an irrelevant velocity. The velocity that *is* relevant in the present theory is the velocity of a charge with respect to the local force field it is traversing. And that velocity, being the (vector) difference between the velocity of the charge and that of the traversed field in the observer's coordinates is *Galilei*-invariant.

In other words, when the velocities in the Maxwell equations and Lorentz force are properly attributed, the equations remain valid in all inertial frames moving in unreformed space and time, and related to each other by the Galileian transformation.

This will be shown in more detail below. However, we must first discuss a point that applies to any velocity, not necessarily high, but is often overlooked: the only measurable and observable quantity in all of electromagnetics is force, usually expressed as the Lorentz force

$$\mathbf{F} = q(\mathbf{E} + \mathbf{v} \times \mathbf{B}) \tag{1}$$

and that includes forces such as those activating the electrons in an antenna and those in the cones of our retinas. Without a charge to multiply them and thus to convert them into a force, the vectors **E, B, D, H, P, M** are abstract aids to our imagination whose existence is *inherently* unprovable. Force is that which changes the momentum of matter, charged or not, such as the pointer of a measuring instrument. There is no way of measuring or demonstrating any quantity without first converting it to force. Therefore as long as force transforms correctly — in accordance with experience — it does not matter that the abstractions expressing it, such as the field vectors by themselves, have certain properties, for example, invariance to a certain transformation. In particular, if the hypothesis that the effect-producing velocity — which means the force-producing velocity — is that of a charge with respect to the local field, then the Maxwell equations will be no less invariant to the Lorentz tranformation with observer-defined velocities than before, but the only thing that matters, the resulting force, will be Galilei-invariant, because it depends on the *difference* between the two velocities, and this relative velocity is conserved.

Referring velocities to a different *type of* standard will, of course, produce a different Lorentz force in certain circumstances, and the difference is in principle measurable. The fact that the difference has not been detected is not due to alternative interpretations of the same evidence, but due to the lack of technology to detect the tiny difference under the circumstances — at least, as yet.

As we have seen, the Lorentz transformation and the Einstein theory, which is based on it, cater to the modification of charge distribution and of the resulting electromagnetic field brought about by the motion of charges. This modification is predicted by the Maxwell equations, and there is no quarrel about its existence, only about its physical origin — that is, about the rest standard to which the velocity producing this effect should be referred. It is therefore clear that differences must be looked for in the interaction of electric charges (or gravitational masses), in particular, in the simplest possible interaction at a distance — that of two charged particles described by the Newton-Coulomb Law.

We shall examine this question in a moment. However, it is also instructive to compare the two theories in cases that do *not* involve any action at a distance, that is, no electromagnetic or gravitational fields. These are also the cases where the Einstein theory is never applied, in part because the low velocities usually involved do not justify its application, but doubtlessly also because these cases reveal the staggering complexity of simple problems in space-time coordinates distorted to cater to something totally unconnected with the problem.

Consider, for example, the case of a windmill used to pump water. The power delivered by the windmill is known to be proportional to the cube of the wind velocity *relative to the windmill*, and to be quite unconnected with electromagnetism. We wish to describe the process in the coordinates of an observer with a general velocity $\mathbf{v}_0$ with respect to the windmill.

Let the wind velocity in the mill rest frame be $\mathbf{v}$; then in the present theory, for an observer going past the windmill at 99% of the speed of light, the wind blows with a velocity $\mathbf{v}+0.99\mathbf{c}$, and the velocity of the windmill in the observer's frame is $0.99\mathbf{c}$. But the velocity producing the power is simply the difference between the wind and the mill velocities, which is $\mathbf{v}$, for in the Galileian transformation the velocity of the observer always cancels in the difference, and the power/velocity function remains unchanged.

In the Einstein theory, if the wind blows at an angle that will give both the wind and the blades the benefit of the Lorentz transformation, the observer sees the wind slowing and the blades shrinking; but the Lorentz-transformed force of the *slower* wind (moving *denser* and *heavier* air due to length contraction and mass increase) on the *narrower* blades acts in dilated time, *lengthening* it. On the pump side, the water has become *heavier*, for its weight mg is affected by the vertical component of the observer's velocity (and as a matter of fact, by the other components as well); it has also become *denser* because the water column has contracted (mercifully, this time due to the vertical component of the observer's velocity only); and the upward velocity of the water will be *slowed* to yield an altered power when multiplied by a modified force. For a given height above the water table, we can ordinarily express the power in gallons of water per second; but in the Einstein theory, this becomes shrunk gallons of a weird liquid per dilated second.

Now all of these stunningly complicated acrobatics are necessary to save a relativity principle (of sorts) catering to the field of a moving electric charge, which has nothing whatever to do with a windmill pumping water.

No such problems arise in a theory that does not refer (all) velocities to an observer: the force of the wind on the blades is the same for all observers, moving or not, as is the power converted.

Since the effect-producing velocities of electric charges are relative to the fields they traverse, not relative to any observers, electromagnetic forces must remain Galilei-invariant for the same reason as the force of the wind on a mill. This will be shown next.

1.10.4. Invariance of the Maxwell Equations

Before we deal with the invariance of the Maxwell equations to the *Galilei* transformation once their velocities have been properly assigned, we must once more go back to the analogy of the wind moving a windmill.

When an observer measuring the windspeed is at rest with respect to the mill (as he usually is), it does not matter whether he refers the wind velocity to himself or to the mill; in either case the windmill equation $W = Cv^3$ linking the power W to the wind velocity v (with C the constant of proportionality) will be confirmed by all measurements.

Now suppose that the observer *falsely* concludes that the effect-producing velocity under all conditions is the velocity referred to himself, but that he *correctly* applies the relativity principle to predict what will happen when he moves with respect to the windmill. Then he will say: it matters not whether the air moves against me or I move against the air; therefore if I start running on a windless day, the windmill must begin to turn. He tries the experiment (which, alas, is far easier performed than its analogy of an observer moving fast through an electromagnetic field) and finds the prediction wrong. If he does not want to sacrifice the relativity principle, he has two choices: he can abandon the false premise, or he can keep it alive by deforming space and time in a transformation of coordinates that will properly cater to the false premise. For example, the transformation might confirm the prediction that his running causes a force by the air on the windmill blades; but it would also cause his running to produce an eqally large torque on the shaft in the opposite direction. That is why runners don't move windmills. (Is that taking the spoof too far? I will remind you shortly, if you think so.)

But this is not a facetious spoof; it is a fairly good analogy of relativity applied to electromagnetics, with the windmill representing a charge and the wind representing the field of another charge. The reason why the Maxwell equations "are not invariant to the Galilei transformation" is that the velocities implicitly occurring in them have been mistakenly referred to the observer as the rest standard. As soon as these velocities are recognized as velocities of charges with respect to the fields they traverse, the Galileian transformation will work on the Maxwell equations as surely as it does on the wind and the windmill.

Before showing that in detail, let me go one step further than saying the field vectors are abstractions which cannot manifest themselves until multiplied by charge to yield an observable force. I will now say that the field of a single charge, unrelated to any other charge, is a meaningless pattern of arrows and rings drawn on papers and blackboards.

This may sound radical, but we need only think back to the definition of a line of *force*, with which the electric field intensity is often plotted. It has the direction in which a small test charge, that is, *a second* charge or a charge *other than* that producing the field, would move if it were present at that point; and the density of the lines is proportional to the magnitude of the force that would act, again, on this test charge. Clearly a field without this second charge is the same type of entity as a waterfall without water.

This is not hairsplitting or nitpicking, for the field of a point charge changes very decidedly when it is in motion, rather than at rest, with respect to this test charge, even though the Einsteinian observer thinks *he* is causing the change with his own motion. (He fails to ask about the velocity of the test charge, just as the analogous observer failed to ask about the velocity of the windmill in his own coordinates.)

Now consider the Maxwell equations. Since they are linear, it is permissible to consider the case of only two moving charges; the case of many charges, up to and including a continuous charge distribution, then follows by superposition – at least in principle, though the actual superposition for charges moving in different directions can be quite dificult.

The two directly recognizable velocities in electrodynamics are the velocity of a charge in a magnetic field, which occurs in the Lorentz force, and the velocity of charges forming a current, which occurs in the current density $\mathbf{J} = \varrho \mathbf{v}$ involved in the second Maxwell equation. To interpret them as "effect-producing" velocities, we must understand them to mean the difference between the velocity of the charge and that of the field which it is traversing – just as a runner should take the difference between the wind and mill velocities that he sees in his own coordinates.

Let an observer move with respect to at least one of the two charges with velocity $\mathbf{v}_0$, and let him describe all phenomena in a coordinate system in which he is the origin. All time derivatives in the Maxwell equations will remain unchanged, for there is no time dilation, and all space derivatives will remain unchanged because there is no length contraction, so that all curls and divergences remain unchanged, too. His velocity is added to all charges and all fields in the same way, so that the difference remains unaltered and his own velocity, which cancels, becomes irrelevant. When it is understood which velocities are effect-producing, the Maxwell equations are as invariant to the Galileian transformation as the wind driving a mill.

Note that a charge density moving with respect to the observer's coordinates, but not with respect to the local field, does not become a current density — at least not one that produces any electromagnetic effect such as a magnetic field. This is analogous to the runner on a windless day: he certainly feels a wind, but not one that will turn a windmill.

That takes care of all observers, who are thus condemned to observing without interfering. It also takes care of the relativity principle, which is satisfied automatically by relative, effect-producing velocities just as in the case of the windmill.

There will, however, still be the physical effects arising when a charge moves through a field. (Analogy: the power-velocity law of the windmill.) These are physical effects predicted by the Maxwell equations; in the present theory they are the same for all observers in all inertial frames, and they have no particular bearing on the relativity principle, which they automatically satisfy.

In part, this point has already been dealt with in Sec. 1.6, where we obtained Poisson's equation for the potential ϕ in the form

$$(1-\beta^2)\frac{\partial^2\phi}{\partial x^2}+\frac{\partial^2\phi}{\partial y^2}+\frac{\partial^2\phi}{\partial z^2}=-\frac{\rho}{\epsilon_0} \tag{1}$$

implying that the concentric equipotential spheres about a moving charge flatten into ellipsoids. The Einstein theory attributes this to length contraction seen only by some observers; but the present theory must interpret this as a genuine effect visible to *all* observers whenever a charge traverses a force field. The reason, for the time being, is that the Maxwell equations say so; in Part Two we will find a direct physical explanation giving more insight.

This flattening of the equipotentials was derived in Sec. 1.6 by eliminating the time derivative in the wave equation by means of the relation

$$\frac{\partial f}{\partial t}=-\mathbf{v}\cdot\nabla f \tag{2}$$

where f is any field component of the moving charge, "moving" meaning with respect to the local field, to the observer or to the ether depending on the corresponding theory. [Relation (2), incidentally, can be derived more cleanly than has been done in Sec. 1.6 by direct use of the Galileian transformation.] However, there is another way of deriving this flattening of the equipotentials, found some 17 years before the appearance of the Einstein theory by the unique genius Oliver Heaviside (1850-1925).[1]

Consider two charges at a rigidly fixed distance from each other, both traveling with velocity **v** with respect to an observer (whom we will also put at rest with respect to the ether). In our theory, this velocity is irrelevant, for one charge is at rest in the field of the other; therefore the force between the two charges will be given by the *static* Coulomb Law and cannot be changed by the motion of an observer or travel through the ether.

As in any other theory, it is of course assumed that the gravitational field is negligibly small compared with the electric field of the charges. This is *not* inconsistent with the explanation of a magnetic field produced by a wire-bound current (Sec. 1.4) or the explanation of the Ives-Stillwell experiment (Sec. 1.9.3), for in both of these cases the electric field or its consequences are absent, leaving nothing but the gravitational field as the "remainder" field. In the case of a wirebound current, the electric field is neutralized by the positive iron grid; in the case of radiating hydrogen atoms, the electric force effectively disappears by averaging, since the crucial point is the effect of the gravitational field *in addition to* the electric field, which is present in both moving and stationary atoms and disappears by subtraction in the comparison.

However, Heaviside [1888, 1889], like all scientists of the time, referred the velocities of charges to the ether, so that the force between the two co-traveling charges was

$$\mathbf{F} = q(\mathbf{E} + \mathbf{v} \times \mathbf{B}) \tag{3}$$

with **v** the velocity of traversing the ether. Since all quantities here are constant in time, **E** has no curl, and is therefore a pure Coulomb field given as the gradient of a potential ϕ. On the other hand, we have, by definition or derivation,

$$\mathbf{B} = -\frac{\mathbf{v} \times \nabla\phi}{c^2} \tag{4}$$

Hence

$$\mathbf{F} = -q\left[\nabla\phi + \frac{\mathbf{v} \times (\mathbf{v} \times \nabla\phi)}{c^2}\right] = -q(1 - \beta^2)\,\nabla\phi \tag{5}$$

[1] The man who predicted the ionosphere, invented the Operational Calculus ("The proof is performed in the laboratory" — 25 years ahead of the mathematicians who found the proof in the Laplace transformation), derived Poynting's Theorem independently of Poynting, invented and masterfully used the delta function, pioneered radio engineering, and has many astounding discoveries to his credit, had no college education. This is reminiscent of Michael Faraday, who had virtually no education at all.

But this can be written as

$$\mathbf{F} = -q\,\nabla\Psi \tag{6}$$

where

$$\Psi = (1 - \beta^2)\phi \tag{7}$$

is Heaviside's "convection potential," again showing the concentric, spherical equipotentials of the potential ϕ of a charge at rest flattening into Heaviside ellipsoids when the charge is moving.

If we direct a rectangular system of coordinates with the x axis along the velocity, the equipotentials of (7) are

$$x^2 + (1 - \beta^2)(y^2 + z^2) = \text{const} \tag{8}$$

The force exerted by the moving charges on each other, by definition of a gradient, is perpendicular to the surface of the equipotential ellipsoid at any point. Hence the force between two charges separated by a rigid distance and moving parallel to the x axis through the ether will in general not be directed along the line joining the two charges: it will deviate from it as shown by the figure. To see this, imagine one of two like charges at opposite ends of a bar passing through the origin, and the other on the Heaviside ellipsoid: the force will be perpendicular to the ellipse, so that except for the four points at the axial intercepts it will not be directed toward the other charge at the origin.

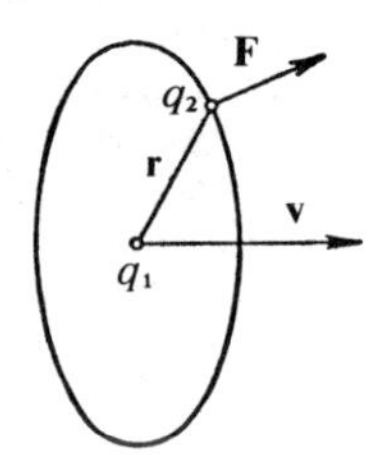

Mutual repulsion of two moving charges.

But it is also evident from the figure that the bar would be subject to a torque. For point charges this torque can easily be calculated by treating them as currents ($qv\,ds$) and using the Biot-Savart Law. For the charges on a parallel-plate capacitor, the result differs only by a factor of ½; the torque seeking to align the plates perpendicular to the "ether wind" due to the translational (orbital) velocity of the earth should be

$$T = \frac{k}{2r}\beta^2 \cos 2\theta \tag{9}$$

where r is the separation of the capacitor plates and θ is the angle between the plates and the translational (orbital) velocity of the earth round the sun. The latter makes $\beta = 10^{-4}$, so that (9) is sufficiently large to be measured if a charged capacitor is suspended from a torsion balance. This was tried in the famous experiment by Trouton and Noble [1903, 1904], but no torque was detected. Their experiment was the electromagnetic equivalent of the Michelson-Morley experiment in that it was also a second-order experiment in β, and in that it also refuted the ether theory, at least in its unentrained version.

Einstein's explanation was simple and similar to the one in the present theory: there is no ether; the velocity of the two charges with respect to the observer (with respect to each other, in the present theory) is zero; nothing is moving, so there is no torque.

But now imagine that, contrary to the Trouton-Noble experiment, the two co-traveling charges move with velocity $\mathbf{v}$ *with respect to an observer.* Then nothing changes in the explanation by the present theory: the velocity of one charge with respect to the other is still zero, and the velocity $\mathbf{v}$ is irrelevant. There is no more torque than in the previous case.

But what does the Einstein theory say?

As long as we consider only electromagnetic forces, as we do here, there can be no difference between the ether theory and the Einstein theory when the ether is replaced by the observer as a standard of rest. This must be true in general, and is easily checked in the present case, for the Heaviside ellipsoids and the Trouton-Noble torque uses nothing but Maxwell's equations and the Lorentz force. To make everything applicable to the Einstein theory, we need only refer $\mathbf{v}$ to an observer rather than to the ether. The Einstein theory must therefore predict exactly the same, non-zero torque as the ether theory.

And it does: the entire calculation can be found in [Becker 1964, pp.397-401]. However, since the moving observer also causes *mechanical* forces to appear, the Einstein theory also predicts a mechanical torque of equal magnitude but opposite direction to which the bar is subject: contraction of the bar in the direction of the velocity is equivalent to a rotation of the bar, so that the electric force on the charges shifts back into the direction of the bar after all. (And this is also the point where I will remind readers, as I promised on p. 93, of their possible dismay that I carried the spoof with the windmill too far.)

So once again the Einstein theory scrapes through by the last twist compensating for all the previous ones, emerging with a result that was obvious from the outset.

In general, the present theory may predict slightly different effects from those predicted by the Einstein theory when the velocity of a charge with respect to the traversed field differs from that with respect to the observer, though the example just discussed shows that this need not *always* be the case.

1.11. Mercury, Mesons, Mössbauer and Miscellaneous

There are a few odds and ends left before the claim of experimentally verified equivalence becomes fully valid.

For example, the advance of Mercury's perihelion should, for tidiness, appear in this "Einstein Minus Zero" part of the book; however, since it involves only gravitation, it will be delayed to Part Three, Sec. 3.2, where readers may be surprised to find that the "Einstein" formula for the advance of Mercury's perihelion was derived by Paul Gerber in 1898, when Albert Einstein was nine years old.

The time dilation allegedly observed on mesons in the atmosphere has been discussed in Sec. 1.6.3. Without a control experiment of the frequency measured in the moving frame, the argument is invalid; but it used to be circular as well, since it used to be based on quantities inferred from the Einstein theory, which they were supposed to prove.

This was particularly drole in the proof that the ratio of the mean free path L *to the energy* W of the mesons is constant. The reason [Tonnelat 1959] is that in the moving system of the mesons the familiar square root $\sqrt{(1 - \beta^2)}$ in the time dilation will cancel against the same square root in the length contraction. But in classical physics the ratio would be just as constant: not because the square roots cancel, but because they were never there in the first place.

The explanation of the Compton effect relies on quantum mechanics, not on the Einstein theory, which is often brought in quite unnecessarily.[1]

The bending of light in a gravitational field follows immediately from our basic assumption that the velocity of light is constant with respect to the local gravitational field in which it propagates. If the field is inhomogenous, then by Fermat's principle it must bend: the fact that it bends towards the denser field implies that light propagates more slowly in denser gravitational fields, just as it does in denser material media.

Quite similarly, electromagnetic waves should propagate slightly more slowly at higher altitudes above the earth, where the gravitational field is less intense. A sufficiently precise standard of a radiated frequency should therefore have a slightly different wavelength – slightly longer at higher altitudes, and a slightly different Doppler shift (which is a function of the velocity of propagation) if the source is moving. Regular sources, including lasers, have too broad a spectrum to detect such minute differences, but the Mössbauer effect, observed on gamma rays emitted in the radioactive decay of certain isotopes at precise energy levels, can be used to detect such differences in Doppler shift for height differences as small as

[1] G. Joos, *Theoretical Physics* (Blackie, London 1947) is apolegetic about deriving it without using the Einstein theory.

12 m. Experiments by Pound, Rebka and others from 1960 onward confirmed this effect predicted by the general Einstein theory, and not surprisingly, this was regarded as another confirmation of the theory.

However, the behavior of electromagnetic waves in a gravitational field must of necessity depend on their velocity of propagation as a function of the field intensity, or to use optical terminology, on the refractive index of a gravitational field. This is the case, as it must be, for Einstein's theory of gravitation, but the corresponding function is arrived at by a physically opaque matrix calculus in non-Euclidean space. On the other hand, for a theory based on the fundamental assumption that the velocity of light is constant wih respect to the gravitational field through which it propagates, the corresponding index of refraction is the first thing the theory must quite naturally ask for and measure (Einstein's coefficients, too, are anchored in experimental measurements). Once the dependence represented by the index of refraction is established, there *cannot* be a difference, as far as I can see, in experimental results involving ray paths and velocities of electromagnetic waves in gravitational fields, no matter how wildly the theoretical interpretations may differ.

There remain exotic phenomena such as black holes, the pulsations of quasars, and the red shift (Einstein's interpretation of the latter is still controversial even among well-bred physicists); and at the other end of the scale, various bewildering behavior by the particles in the subnuclear zoo. Both of these fields are subjects that I do not propose to enter, so with my inborn modesty I will restrict my claims of equivalence to the narrow domain between the atomic nucleus and the outer reaches of the solar system.

Part Two

Einstein Plus One

2.1. Strictly Central Motion

The following section reviews elementary celestial mechanics; its results will often be referred to in the following.

A central vector field is one whose lines of force are everywhere directed toward (or away from) the same point; the scalar value need not necessarily be governed by the inverse square law, but it is assumed to hold here. In polar coordinates r, θ, with origin in the center of attraction, and unit vectors $\mathbf{r}_0$, $\boldsymbol{\Theta}_0$:

$$m\ddot{\mathbf{r}} = \frac{K}{r^2}\mathbf{r}_0 \tag{1}$$

where $\mathbf{r}$ is the position vector of the attracted body, dots denote differentiation with respect to time, m is the mass of the body or particle acted on, and K is a constant that in Part Three of this book will assume the value corresponding to a gravitational field, but in this entire Part Two, it will have the value corresponding to Coulomb's Law, in particular, to an electron in the Coulomb field of the nucleus. In the latter case

$$K = -\frac{q^2}{4\pi\epsilon_0} = -2.3 \times 10^{-28}\ \mathrm{N/m^2} \tag{2}$$

where q is the electron charge and ϵ_0 the permittivity of free space. Since the nucleus is roughly 1,800 times heavier than the electron, we will not bother about reduced mass, but will assume that the center of the nucleus is also the center of the field.

Equation (1) is an expression on which all theories agree when the velocity $\mathbf{v}_t$ of the attracted particle is small compared with the velocity of light. In Einstein's theory this approximation would involve the neglect of the variability of mass, in the present theory it involves, even to first order in β, the neglect of the θ component on the right; in classical instant-action-at-a-distance (IAAD) physics, (1) is valid as it stands.

Since by (1) the force has the direction of $\mathbf{r}$, we have

$$\mathbf{r} \times m\ddot{\mathbf{r}} = 0 \tag{3}$$

Integrating by parts, we have (for the remaining integral vanishes due to the cross product of $\dot{\mathbf{r}}$ with itself)

$$m\mathbf{r} \times \dot{\mathbf{r}} = m\mathbf{r} \times \mathbf{v} = \text{const} = \mathbf{L} \tag{4}$$

The vector **L** is called the angular momentum and plays a central role in celestial mechanics: in classical physics, i.e., in the force field (1), it is constant for any orbit. However, as can be seen at a glance, (3) will not hold if the motion is not strictly central, and neither will then (4).

The constancy of **L** enables us to separate the variables in the polar equations of motion:

$$m(\ddot{r} - r\dot{\theta}^2) = \frac{K}{r^2} \tag{5}$$

$$mr^2\dot{\theta} = L \tag{6}$$

where (6) is a re-write of (4), and the parenthesis in (5) is the acceleration in polar coordinates.

Eliminating $\dot{\theta}$ from (5) and (6) yields

$$m\ddot{r} = \frac{L^2}{mr^3} + \frac{K}{r^2} \tag{7}$$

This equation will be integrated over r, not time; we therefore use

$$\ddot{r} = \dot{r}\frac{d\dot{r}}{dr} = \frac{d(\frac{1}{2}\dot{r}^2)}{dr} \tag{8}$$

The integration over r is then straightforward, and we have

$$\tfrac{1}{2}m\dot{r}^2 + \frac{L^2}{2mr^2} - \frac{K}{r} = E \tag{9}$$

where the constant of integration E is obviously the total energy, made up of the kinetic energy of the radial component, the kinetic energy of the θ component, and the potential energy. As before in polar coordinates, v_t is the total velocity, leaving the unsubscripted symbol v for the more important velocity in the θ direction:

$$\mathbf{v}_t = \dot{r}\mathbf{r}_0 + v\mathbf{\Theta}_0 \tag{10}$$

The energy equation (9), even before a second integration yields the orbit, provides two important relations. If a is the maximum distance from the center (the semimajor axis of the ellipse that we are about to obtain), the total energy of the electron is from (9), after using (7) with $r = a$,

$$E = -\frac{K}{2a} \tag{11}$$

and if we substitute this value of E in (9) (which is, of course, valid for *any* point of the path), noting that the sum of its first two terms equals the kinetic energy, we find the velocity of the particle or planet at any point of its path:

$$v_t^2 = \frac{K}{m}\left(\frac{2}{r} - \frac{1}{a}\right) \tag{12}$$

To find the orbit $r(\theta)$ we integrate (9) after changing the independent variable from t to θ using (6), i.e. $dv=(L/mr^2)dt$, obtaining

$$\theta = \int \frac{L\,dr}{r^2\sqrt{2m(E+K/r)-L^2/r^2}} \tag{13}$$

To integrate this, set $r=1/u$,

$$\theta = -\int \frac{L\,du}{\sqrt{2m(E+Ku)-L^2u^2}} \tag{14}$$

and complete the square under the square root, showing the integral to be an arc cosine. Reverting from u to r we find the orbit in the form

$$r = \frac{p}{1-\epsilon\cos\theta} \tag{15}$$

which is the equation of an ellipse with one of its foci at the origin. The parameter

$$p = \frac{L^2}{Km} \tag{16}$$

is the semilatus rectum, and

$$\epsilon = \sqrt{1-\frac{L^2}{mKa}} \tag{17}$$

the eccentricity (see figure for a definition of the last two).

The time T taken for one revolution can be found from the area of an ellipse, πab or $\pi a\sqrt{(1-\epsilon^2)}$, and from (6), which contains the areal velocity $\frac{1}{2}r^2\dot{\theta}$ of the position vector $\mathbf{r}$, since an elementary triangular area is $\frac{1}{2}r^2\,d\theta$; the area swept out by the radius vector is shaded in the figure. If the period thus found is is T, then for the frequency ω we have

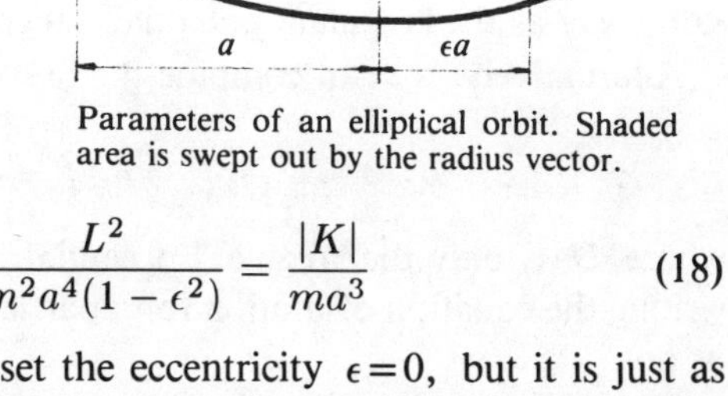

Parameters of an elliptical orbit. Shaded area is swept out by the radius vector.

$$\omega^2 = \left(\frac{2\pi}{T}\right)^2 = \frac{L^2}{m^2a^4(1-\epsilon^2)} = \frac{|K|}{ma^3} \tag{18}$$

For a circular orbit, we can simply set the eccentricity $\epsilon=0$, but it is just as quick and gives more insight to start from the equilibrium equation that expresses the balance of attractive and centrifugal force:

$$\frac{mv^2}{r} = \frac{|K|}{r^2} \tag{19}$$

whence

$$v^2 = \frac{|K|}{mr} \tag{20}$$

For a circular orbit we can also use (19) to express the angular momentum, defined by (4), as

$$L = \sqrt{Kmr} \tag{21}$$

showing that the angular momentum increases with increasing radius (because the radius increases faster than the velocity decreases).

Now let us go back to (9). We obtained this simply as the first integral of the original equation of motion, but the constant of integration E is obviously the total energy, made up of the kinetic energy associated with the radial velocity $\dot{r}$, the kinetic energy associated with the θ component of the velocity, and the potential energy K/r. There is only one derivative in this equation, making its integration by separating the variables r and t trivial; however, it can be written quite generally as

$$T = \tfrac{1}{2}m\dot{r}^2 = E - U \tag{22}$$

where

$$U = \frac{L^2}{2mr^2} - \frac{K}{r} \tag{23}$$

Equation (22) gives us general instructions how to integrate, and it also has a simple physical meaning if the motion is rectilinear: kinetic energy equals total minus potential energy. This simple physical meaning can be conserved for curvilinear motion if we call (23) the "effective" potential energy, the first term being the "centrifugal energy" added to the regular potential energy. This is not altogether artificial, because it is derived by integrating the centrifugal force L^2/mr^3 in the same way as the "regular" potential integrates the Newton-Coulomb force.

Alternatively, we can combine the first two terms in (9) into

$$\tfrac{1}{2}mv_t^2 = E - U \tag{24}$$

where U is only the "regular" potential. In either case, this is a general way of writing the equation of motion for a particle, whether it is moving in a central field or not.

Finally, we note that the potential energy of a body in circular orbit in a central, inverse square-law field is

$$U = \frac{K}{r} = -2\pi L\nu \tag{25}$$

where $\nu = \omega/(2\pi)$ is the orbital frequency (orbits per second) and ω is the angular frequency (in radians per second), which equals v/r. This follows immediately from the equilibrium equation (19), which requires

$$-\frac{K}{r} = mv^2 = \frac{Lv}{r} = L\omega \tag{26}$$

2.2 Self-Induced Oscillations of an Accelerated Charge

In the case of the collective flow of electric charges, no one doubts the phenomenon of self-inductance: as the current increases, a magnetic field builds up, and this change in the magnetic field induces an electric field which opposes the increase in current.

"As the current increases" refers to either more charges flowing or the same charges being accelerated. In the case of a single moving charge, such as an electron, only acceleration is involved, but the same phenomenon must set in. In the macroscopic case, we may produce the oscillations of a tuned circuit if we supply a capacity, storing the energy of the collapsing magnetic field as an electric field, and vice versa. But this capacity merely *increases* the storage for that energy; if it were absent, some electrical energy would still be stored as an electric field in space. Quite similarly, in the case of an electron, we need not supply the energy storer: the electric energy of the Faraday (induced) field is stored automatically in the space surrounding the electron and is converted to and from magnetic energy as the electron accelerates and decelerates about a mean velocity. All a material capacity does is bring down the natural frequency of the oscillations from some 10^{16} Hz to lower values.

The quantitative behavior of this oscillatory process will be derived in later sections. In this overview section, however, let us only consider what must be true for single electron no less than for a group of electrons (a current):

When an electron (not necessarily in orbit) is accelerated, it must, by Ampère's Law, build up a magnetic field. That build-up, by Faraday's Law, will induce an electric field, whose direction is by Lenz's Law opposed to the acceleration. That is, the electron's acceleration is the cause of its *de*celeration. The latter will reduce the magnetic field and induce a Faraday electric field that accelerates the electron again.

Thus, we may expect an electron to accelerate and decelerate from its mean velocity in an oscillatory, periodic motion. The phenomenon is one of self-inductance, with the oscillations reminiscent of a tuned circuit: the energy storage in a capacitor and inductor is replaced by the energy storage in the electric and magnetic field of the moving electron.

The existence of such oscillations will be demonstrated in the next section, where it will be derived directly from the electromagnetic potentials.

Now consider the energy of an electron undergoing such self-inductive velocity-oscillations, in particular, an electron in orbit about the nucleus. Contrary to the central motion discussed in the preceding section, the electron's strictly mechanical energy is *not* constant: there is a periodic flow of energy from the kinetic energy of the electron to the surrounding electromagnetic field and back again – a spatial flow from the electron to its neighborhood, quite unlike the conversion of kinetic to potential energy, both of which remain tied strictly to the electron (or planet) itself. The *total* energy, mechanical plus electromagnetic, must of course remain constant.

In a classical IAAD orbit, the kinetic energy of an orbiting body need only be balanced by the potential energy in such a way that the two add up to a constant. *Any* constant is possible, each corresponding to a family of conics, but if we limit the following to circles, each constant corresponds to a possible radius. That is why IAAD mechanics allows orbits at *any* distance from the center of attraction.

But when the kinetic energy of a charged body fluctuates because a part of it is converted to electromagnetic energy and back, it must, as we shall see, fulfill an additional condition: the natural frequency of these oscillations (to be discussed in later sections) must be an integral multiple of the orbital frequency, or the orbit will not be stable. It is this additional condition that allows only a discrete set of orbits: Bohr's orbits, as we shall see.

Although we are not yet ready to discuss this natural frequency, there is an important relation binding it to the energy and the mean velocity of the electron. This relation can be found from elementary considerations, which are not limited to electrons in orbit, but apply to any charged body moving along an arbitrary path, for example, a straight line.

If the frequency of the velocity oscillations is ν, the average velocity (about which the velocity fluctuates) is v, and the distance (measured along the path) between the points at which the electron attains successive maxima of its fluctuating velocity is λ, then by elementary kinematics these three quantities must be related by

$$v = \nu\lambda \tag{1}$$

In spite of its formal appearance, (1) has nothing to do with radiation or a propagating wave; it is a trivial consequence of elementary kinematics, not radically different from that applying to a cook chopping a carrot (ν chopping frequency, λ slice thickness, v velocity of feeding the carrot to the knife). For this reason I will at times refer to this basic relation as the "carrot formula."

Now let us consider some other basic laws which we will need in the following.

The purely mechanical kinetic energy of a charged body, which it would have even if it were not charged, is found in the usual way by integrating the work done by the force accelerating the body from 0 to v:

$$T_{kin} = m\int_0^s \frac{dv}{dt}ds = \int_o^v v\,dv = \tfrac{1}{2}mv^2 \tag{2}$$

But if the body is charged, its motion at velocity v carries additional energy, for the force accelerating it from rest to velocity v now has to do work not only against mechanical inertia as above, but also against the electric field induced by the magnetic field that increases as the charge accelerates. In other words, the force has to overcome not only the inertial reaction due to Newton's Second Law, but also the electromagnetic reaction due to Faraday's Law.

This additional work done against the Faraday reaction in accelerating the charge to velocity v appears as the energy stored in the magnetic field. It can therefore be calculated by integrating the magnetic energy density over all space outside the spherical charge with radius R moving with velocity v. The magnetic energy density of a moving charge is found by expressing **B** by the Coulomb field **E** as in (3), Sec. 1.5,

$$\frac{B^2}{2\mu} = \frac{v^2 E^2 \sin^2\theta}{2\mu c^4} = \frac{\mu q^2 v^2 \sin^2\theta}{2(4\pi)^2 r^2} \tag{3}$$

where r is the distance from the charge, and where we have used $\mu\epsilon = 1/c^2$.

This is integrated with an element of volume

$$dV = 2\pi r^2 \sin\theta \, dr \, d\theta \tag{4}$$

over all space outside the sphere with radius R, i.e. over θ from 0 to π and over r from R to ∞. An elementary integration then yields

$$T_{mag} = \frac{\mu q^2 v^2}{12\pi R} \tag{5}$$

where R is the electron radius (or for that matter, the radius of a moving charged tennis ball, but the result will show why the magnetic energy of charged tennis balls is not very impressive).

Comparing this with (2), we see that the quantity

$$m_{mag} = \frac{\mu q^2}{6\pi R} \tag{6}$$

plays the role of an inertial mass. This expression agrees with (12), Sec. 1.5, where we obtained it as an approximation for slow charges by considering the electromagnetic momentum of a moving charge.

It will be seen that this electromagnetic mass depends not only on the charge, but also on the radius R; it increases as the charge is concentrated into a smaller volume.

The numerical value of R of an electron, which has never been directly measured, will not be important in the following, nor can we calculate it yet, since we do not know the value of m_{mag}. But we may note in passing that if we assume the latter of the same order as the electron mass m, then on substituting the electron charge and mass in (6), we obtain the approximate value

$$R_{mag} = \frac{\mu q^2}{6\pi m_{mag}} = 1.88 \times 10^{-15}\ \mathrm{m} \tag{7}$$

which is of the same order as other values derived for R in classical physics, e.g. by integrating the electrostatic energy of an electron over space (as we have done here for the magnetic energy) and setting the result equal to mc^2. Various other methods also yield values of the same order.[1]

However, more important than the numerical value of R is the relation between the kinetic energy (2) and the electromagnetic energy (5), which depends on the ratio of electromagnetic mass to Newtonian (uncharged) mass. For a charged tennis ball with known radius and Newtonian mass, the ratio is easy to determine (it is very close to zero). For an electron, we do not know the corresponding values, but for reasons that will be apparent in a moment, we will assume that the two masses, and therefore also the two energies, are equal. The total energy of an electron moving with velocity v is then

$$T = T_{kin} + T_{mag} = 2T_{kin} = mv^2 = mv\lambda\nu \tag{8}$$

where (1) has been used in the last equality.

For an electron under given constraints, there will be a natural fundamental frequency plus harmonics as in any other system, and we shall later find them for the case of an electron in orbit. However, what we can say already from (8) is that for a given ν the first three factors on the right must be constant, since the total energy T is constant. Hence

$$mv\lambda = h \tag{9}$$

where h is a constant – its value as yet undetermined.

Relation (9) is the reason for the assumption of equal masses. For (9) is the well tested de Broglie relation, introduced in 1924 as an independent postulate to explain the Bohr orbits of an electron, and in turn leading to the Schrödinger equation and quantum mechanics. As yet, the identity of (9) and the de Broglie relation is purely formal, for in the latter, λ is the de Broglie wavelength and h is Planck's constant, whereas in (9), λ is the distance between two successive velocity maxima along the electron's path, and h is a mere constant of proportionality. We shall, however, use (9) to derive the Bohr orbits, and since the resulting formulas have h in the usual places, it will follow that h must be Planck's constant, which by (9) will then also fix λ as the de Broglie wavelength.

In the meantime, please note that the fundamental relation (9) and the constant h involved in it have been derived here without new assumptions. (The assumption of equal electron masses only fixes the numerical value in accordance with experiment; it is unnecessary for the functional form of the relation.) In particular, the constant h appears in (9) as a result of only the "carrot formula" (1), and without the use of black-body radiation or atomic spectra.

[1] The value (7) agrees with that derived by Barnes [1983], who regards magnetic and kinetic energy not just as equal, but as identical. See also Sec. 2.8 for a discussion of the value R.

2.3. The Faraday Field and Electron Velocity Oscillations

The electric field of a charge is usually treated as a single physical quantity, described by a single symbol **E**. Yet in fact it is the sum of two fields, both of which are associated with the force on a charge, but which are otherwise quite different: one is the irrotational Coulomb field, which attracts or repels other charges in a way reminiscent of gravitational force; the other, with non-zero curl, is the induced Faraday field associated with acceleration and reminiscent of inertial reaction.

The Coulomb field, directed radially outward from a point charge, is often misleadingly labeled "electrostatic," though it is time-varying for an observer moving through it, which by the relativity principle is equivalent to a moving charge carrying the Coulomb field with it.

Let us therefore write

$$\mathbf{E} = \mathbf{E}_c + \boldsymbol{\psi} \tag{1}$$

where $\mathbf{E}_c$ is the Coulomb field and $\boldsymbol{\psi}$ is the Faraday field. What distinguishes the two fields is the absence or presence of a curl:

$$\nabla \times \mathbf{E}_c = 0 \tag{2}$$

whereas on substituting (1) and (2) in the first Maxwell equation we have

$$\nabla \times \boldsymbol{\psi} = -\frac{\partial \mathbf{B}}{\partial t} \tag{3}$$

The two fields are most easily analyzed by using the electromagnetic potentials. Since the divergence of a curl vanishes identically, it follows from the Maxwell equation $\operatorname{div}\mathbf{B}=0$ that $\mathbf{B}$ is derivable from a vector potential $\mathbf{A}$ through

$$\mathbf{B} = \nabla \times \mathbf{A} \tag{4}$$

Similarly, since a gradient has no curl, it follows from (2) that the Coulomb field is derivable from a scalar potential ϕ,

$$\mathbf{E}_c = -\nabla \phi \tag{5}$$

Eliminating **B** from (3) and (4) yields

$$\boldsymbol{\psi} = -\frac{\partial \mathbf{A}}{\partial t} \tag{6}$$

so that the two fields follow from the potentials by (5) and (6).

The equations for these potentials are found by substituting in (4) using the second Maxwell equation

$$\nabla \times \mathbf{B} - \frac{1}{c^2}\frac{\partial \mathbf{E}}{\partial t} = \mu \mathbf{J} \tag{7}$$

On using the vector identity for curl curl $\mathbf{A}$ we have

$$\nabla^2 \mathbf{A} - \nabla(\nabla \cdot \mathbf{A}) - \frac{1}{c^2}\frac{\partial \mathbf{E}_c}{\partial t} - \frac{1}{c^2}\frac{\partial \boldsymbol{\psi}}{\partial t} = -\mu \mathbf{J} \tag{8}$$

Only the curl of $\mathbf{A}$ has so far been defined in (4), so that we are still free to choose its divergence. To convert (8) into a tidy wave equation, we choose it (in the so-called "Lorentz gauge") as

$$\nabla \cdot \mathbf{A} = -\frac{1}{c^2}\frac{\partial \phi}{\partial t} \tag{9}$$

so that on using (5) the second and third terms of (8) vanish, and using (6), equation (8) reduces to

$$\nabla^2 \mathbf{A} - \frac{1}{c^2}\frac{\partial^2 \mathbf{A}}{\partial t^2} = -\mu \mathbf{J} \tag{10}$$

To obtain an equation for ϕ we use the Maxwell equation for div $\mathbf{E}$, which by (1), (5), (6) and (9) yields

$$\nabla^2 \phi - \frac{1}{c^2}\frac{\partial^2 \phi}{\partial t^2} = -\frac{\rho}{\epsilon} \tag{11}$$

The solution of (11) will be found in standard textbooks: by transforming the time derivative to a space derivative (see Sec. 1.7.5), one obtains a Poisson equation, whose solution (at least for small v) is

$$\mathbf{A} = \frac{\mu}{4\pi}\iiint \frac{\mathbf{J}}{r}\, dV \tag{12}$$

where $\mathbf{r}$ is the distance from the fixed point of observation, at which $\mathbf{A}$ is to be calculated, to a running point in the volume V containing all currents. On substituting

$$\mathbf{J} = \rho \mathbf{v} \tag{13}$$

where ϱ is the charge density and $\mathbf{v}$ the velocity with which the charge is moving, one finds after a fairly simple integration

$$\mathbf{A} = \frac{\phi \mathbf{v}}{c^2} \tag{14}$$

where ϕ is the electrostatic potential of the charge.

From (13) and (6) the Faraday field is

$$\boldsymbol{\psi} = -\frac{1}{c^2}\left(\phi \frac{d\mathbf{v}}{dt} + \mathbf{v}\frac{d\phi}{dt}\right) \tag{15}$$

where ϕ is the Coulomb potential (possibly delayed) moving with its source charge. The Faraday field at any point is thus induced by two causes, corresponding to the two terms in (15): acceleration, and a Coulomb field moving past that point. Both are accompanied by a change in magnetic field, which is the more usual way of explaining the Faraday field.

The Coulomb potential of a point charge $-K/r$ applies to a static point charge, or by the principle of relativity, to one moving uniformly with respect to the standard of rest. But here we have an *accelerated* charge, and the principle of relativity no longer guarantees that the static field will move "frozen" to the charge; in fact, if the field is formed by propagation from the charge, as we assume, an accelerating charge will catch up with, or lag behind, the equipotentials, i.e., run into a higher gradient on the forward side and leave behind a lower one (see figure). Strictly speaking, we should therefore use the modified Newton Law with the delays as in (19), Sec. 1.8. This will produce a potential valid at a time delayed by r/c after the potential has been "emitted," i.e., produced by the charge where it was r/c seconds ago. The variable delay will produce the non-concentric circles shown in the figure (this corresponds to the Liénard-Wiechert formulas of a delayed potential). However, until Sec. 2.7 we will evade this effect by explicitly substituting an expression for the potential (or field) only in the immediate neighborhood of an accelerated charge, that is, before a significant delay can materialize.

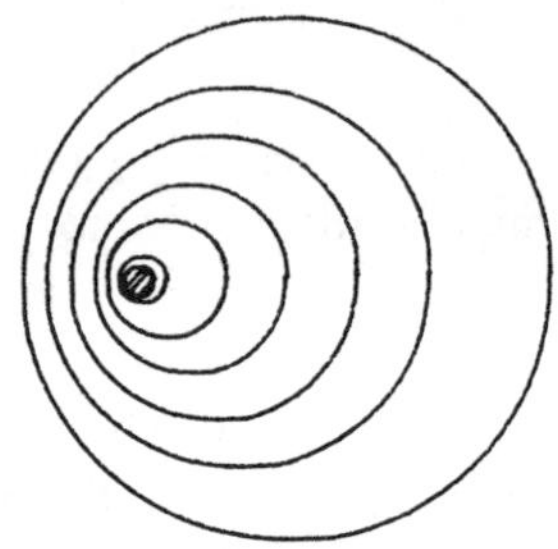

Equipotentials of an accelerated charge

Let us now apply (4) to (14), using the vector identity

$$\nabla \times (\phi \mathbf{v}) = \nabla\phi \times \mathbf{v} + \phi\nabla \times \mathbf{v} \tag{17}$$

If $\operatorname{curl}\mathbf{v}=0$, this immediately yields

$$\mathbf{B} = \frac{\mathbf{v} \times \mathbf{E}_c}{c^2} \tag{18}$$

in agreement with (3), Sec. 1.5, where it was used as a *definition* of the magnetic field. The present derivation of this much used and rarely derived formula shows that it is valid only for the Coulomb field $\mathbf{E}_c$, not for the total $\mathbf{E}$ field given by (1).

We may also note that the divergence of $\boldsymbol{\psi}$ follows from (1) and from $\operatorname{div}\mathbf{E}=\varrho/\epsilon$:

$$\nabla \cdot \boldsymbol{\psi} = 0 \tag{19}$$

That is, the lines of $\boldsymbol{\psi}$ are closed, possibly via infinity.

Now let us return to (15) to show that for the case of a single charge (which we take spherical with uniform surface density) in *its own* Faraday field, the second term is usually of no importance. The force on the charge by its own field is

$$q\boldsymbol{\psi} = \iiint \left[\rho\phi\frac{d\mathbf{v}}{dt} + \rho\mathbf{v}\frac{d\phi}{dt}\right]\, dV \tag{20}$$

Unlike the first term, the second will vanish in the integration, for

$$\frac{d\phi}{dt} = \frac{d\phi}{dr} v \cos\theta, \qquad dV = 2\pi r^2 \sin\theta \, dr \, d\theta$$

so that the integral over θ will vanish due to symmetry – just as in an integration of the force on a charge by its own Coulomb field (to which, in fact, this term is proportional). For the calculation of the Faraday force $q\boldsymbol{\psi}$ on a charge in its own Faraday field, therefore, the second term in (15) is merely waiting to vanish in the inevitable integration, and is as good as non-existent. For the present application we may therefore write

$$\boldsymbol{\psi} = -\frac{\phi}{c^2}\frac{d\mathbf{v}}{dt} \tag{21}$$

If we multiply both sides by q, the left side is the force on an accelerated charge q by its own electric Faraday field $\boldsymbol{\psi}$, and on the right side the quantity $-q\phi/c^2$ has taken the place of the inertial mass m. Hence the energy

$$E(\text{nergy}) = q\phi = mc^2$$

which is one of several ways of obtaining this formula without the Second Postulate and the Lorentz transformation; however, I do not consider any of these methods as clean as the one leading to (13), Sec. 1.7, which does not need the Lorentz transformation, either.

Let us now fix the coordinate origin in the electron. (As we shall see in a moment, the electron velocity oscillates about an average value, so the origin is actually *near* the electron, at the point of its average position.) The vanishing derivative of the Coulomb field in the Maxwell equation (7) will cause $\mathbf{E}$ to be replaced by $\boldsymbol{\psi}$ in that equation; dot multiplying (7) by $\boldsymbol{\psi}$ and (3) by $\mathbf{H} = \mathbf{B}/\mu$, and proceeding in the usual manner to obtain the Poynting-Heaviside Theorem, we have

$$\iint (\boldsymbol{\psi} \times \mathbf{H}) \cdot d\mathbf{S} + \iiint \mathbf{J} \cdot \boldsymbol{\psi} \, dV = -\frac{\partial}{\partial t} \iiint [\tfrac{1}{2}\epsilon\psi^2 + \tfrac{1}{2}\mu H^2] \, dV \tag{22}$$

The right side expresses the change in electromagnetic energy within the volume V. If this energy is to be conserved, the right side must equal zero. The integrand is non-negative, $\boldsymbol{\psi}$ is by (20) proportional to $\dot{\mathbf{v}}$, and $\mathbf{H}$ is by (18) proportional to $\mathbf{v}$; hence

$$\frac{\partial}{\partial t}[C_1 \dot{v}^2 + C_2 v^2] = 0 \tag{23}$$

where both constants are positive.

Performing the differentiation, canceling by $2\dot{v}$ and setting $C_2/C_1=\omega^2$, we have

$$\ddot{v} + \omega^2 v = 0 \tag{24}$$

which is the equation of an oscillating velocity with a sine-cosine solution.

Thus the velocity oscillations which had been expected for qualitative, physical reasons in Sec. 2.2 as a consequence of Faraday's Law of self-induction, have now been shown to exist directly from the electromagnetic potentials.

The velocity in (24) was defined with respect to a system moving with the electron (or more accurately, with respect to a system in which the electron's average velocity is zero). For a system in which the electron moves with average velocity v_0 (about which the velocity v fluctuates), we need only replace v in (24) by $v - v_0$. The solution, if we choose our time origin so that only the sine is retained, is then

$$v = v_0(1 - a \sin \omega t) \tag{25}$$

where a is a constant to be determined in the next section.

From the velocity we can calculate the path $r(\theta)$; for example, for the special case of an electron orbiting closely to the IAAD circle, we might use (19), Sec. 2.1, as an approximation, yielding

$$r = \frac{K}{mv^2} = r_0(1 + 2a \sin \omega t) \tag{26}$$

where r_0 is the IAAD radius.

However, the velocity-distance relation used in (26) assumes strictly central motion, which is not the case when the Faraday force acts on the electron in the transversal direction. A more careful treatment, to be performed in the next section, will yield a different constant, though it does not change the form of (26). For the time being, therefore, we will write it with an as yet undetermined constant as

$$r = r_0(1 + b \sin \omega t) \tag{27}$$

Thus the orbit is one that wiggles about the IIAD circle with radius r_0. The wiggle frequency ω is determined by the two constants in (23), and if we assume that it equals the orbital frequency – a point that will be confirmed in the next section – this will make ω a function of r. Hence the relation (27) is non-linear, and the fundamental frequency ω will have its harmonics, so that the general path of an electron in near-central motion (the transversal Faraday force prevents it from being strictly central) is

$$r = r_0 \left(1 + \frac{b}{n} \sin n\omega t\right) \tag{28}$$

This electron path, slightly wiggling about the circular IAAD orbit, was here derived from electromagnetic considerations. Next, it will be shown that it can also be derived, and with somewhat more insight, by celestial mechanics.

2.4. Slightly Off-Central Motion

Consider a charged particle (electron) moving in a central field, but subject to a small Faraday force $q\boldsymbol{\psi}$, in scalar value equal to γ times the absolute Coulomb force, where γ is an as yet unknown small constant or function. The radial component of this force is negligible compared with the Coulomb force, so that the total force acting on the particle is

$$\mathbf{F} = \frac{K}{r^2}(\mathbf{r}_0 + \gamma\mathbf{\Theta}_0) \tag{1}$$

This is no longer a strictly central force due to the presence of the transversal or θ component. There are two important differences compared with the strictly central motion discussed in Sec. 2.1.

First, the angular momentum L is no longer constant, as immediately apparent from (3) and (4) of Sec. 2.1.

Second, the total energy includes not only kinetic and potential as before, but also the electromagnetic energy by which the Faraday force accelerates and decelerates the orbiting charge. Of this total energy, mechanical plus electromagnetic, part of one is being converted into part of the other: the energy of an accelerating and decelerating electron flows to and from its magnetic field, the acceleration and deceleration being effected by the Faraday field as the magnetic field builds up or collapses. But it is not immediately obvious by how much the energy determining the orbit as in (9), Sec. 2.1, should be increased: although we know the total amount of electromagnetic energy from Sec. 2.2, this is of no help, for some of that energy is interchanged between the magnetic and the Faraday fields in parts of space where it performs no work — it simply builds up a Faraday field far from the charge, and this energy has no effect on the trajectory of the particle.

We therefore set the additional energy equal to the force K/r^2 times an as yet unknown constant distance κ to be determined from the solution.

The equation to be solved to find the electron path for this case is then

$$E = \tfrac{1}{2}m\dot{r}^2 + \frac{L^2}{2mr^2} + \frac{K}{r} - \frac{K}{r^2}\kappa \tag{2}$$

where the angular momentum $L = mvr$ is no longer constant.

Please avoid confusion and note that from now on, whenever polar coordinates are used, v stands for the transversal or θ component of the velocity, not for the total velocity

$$\mathbf{v}_t = \dot{r}\mathbf{r}_0 + v\mathbf{\Theta}_0$$

and $\beta = v/c$ likewise refers to the transversal velocity.

Also, E in this section stands for energy, not electric field strength.

As a matter of fact, (2) can be solved exactly, but the solution is inverted (giving t as a function of r), and in such a messy form that it does not provide immediate insight.

The details, in brief outline, are as follows. By using (12), Sec.2.1, we have $L = hr/m\lambda$, so that (2) becomes

$$\dot{r}^2 = \frac{2E}{m} - \frac{h^2}{m^2\lambda} + \frac{2K}{mr} - \frac{2K\kappa}{r^2 m} \tag{3}$$

On taking the square root, separating the variables and a somewhat lengthy integration, one obtains

$$t = \frac{r}{C}\sqrt{A + \frac{B}{r} - \frac{C}{r^2}} - D\arcsin(F + Gr) \tag{4}$$

where A through G are messy functions of the constants in (3). What we need is the orbit $r(t)$, i.e., the inverse function of (4); this can be found numerically by computer, which (as usual with computers) provides high accuracy, but little insight.

Instead, we will use a perturbation method. Since (1) differs from the IAAD (instant action at a distance) force only by a term involving the small quantity β, the solution may also be expected to be close to the classical one, so that we can represent the solution $r(t)$, and similarly the angular momentum $L(t)$, in the form

$$r(t) = r_0[1 + \epsilon_1(t)] \tag{5}$$

$$L(t) = L_0[1 + \epsilon_2(t)] \tag{6}$$

where L_0 and r_0 are the classical IAAD values, and the absolute values of the correcting functions ϵ_i are small compared with unity throughout the orbit. We can therefore approximate by interchanging the *undifferentiated* variable L and the constant L_0 at will, and similarly so with r and r_0, but of course we cannot take such a cavalier attitude to the derivatives, which may differ considerably in spite of the closeness of the originals. For IAAD, L_0 is a constant always, and we take r_0 constant also, i.e. the unperturbed case of a circle rather than an ellipse. (This restriction may be withdrawn at the cost of substantial, but probably irrelevant, complications.)

The derivative of the no longer constant angular momentum $\mathbf{L}$ can be found as follows:

$$\dot{\mathbf{L}} = m\dot{\mathbf{r}} \times \mathbf{v} + m\mathbf{r} \times \dot{\mathbf{v}} = \mathbf{r} \times \mathbf{F} = \frac{vK}{cr}\mathbf{z}_0 \tag{7}$$

where the first equality differentiates the definition of $\mathbf{L}$; the first term vanishes since it cross-multiplies identical factors, the second contains the definition of force, of which only the $\mathbf{v}$ component survives the cross product, and the last expression substitutes from (1). The unit vector $\mathbf{z}_0 = \mathbf{r}_0 \times \mathbf{\Theta}_0$ shows that the orbit remains in the same plane as in strictly central motion.

However, what we shall need is not the derivative of L by itself, but the ratio $\dot{L}/\dot{r}$, and this can be determined from the conservation of energy (or the balance of powers):

$$\dot{r}F_r = vF_\theta \tag{8}$$

Using (1), we have from (7) and (8)

$$\dot{L} = rF_\theta = \frac{\dot{r}K}{vr} \tag{9}$$

so that by (20), Sec. 2.1,

$$\dot{L} = \frac{\sqrt{Kmr}}{r} \tag{10}$$

or using (21), Sec. 2.1,

$$\frac{\dot{L}}{\dot{r}} = \frac{L}{r} \tag{11}$$

which is the desired ratio with which to handle derivatives of L.

This establishes the tools, and we now return to (2).

First we differentiate it with respect to time, remembering that L is not constant:

$$m\dot{r}\ddot{r} - \frac{L^2\dot{r}}{mr^3} + \frac{L\dot{L}}{mr^2} - \frac{K\dot{r}}{r^2} + \frac{2K\kappa\dot{r}}{r^3} = 0 \tag{12}$$

Dividing both sides by $m\dot{r}r$ and using (11), we find

$$\frac{\ddot{r}}{r} - \frac{K}{mr^3} + \frac{2K\kappa}{mr^4} = 0 \tag{13}$$

Up to this point, to the extent that (9) is (physically) valid, our procedure is (mathematically) still exact. It is only now when the first derivatives of L and r have been eliminated that we use the perturbation approximation. Since we no longer run the danger of ignoring the effect of the first derivatives in (2) or (12), i.e., in the second and third terms of (13), we will now freely interchange r as a constant or a variable in those two terms by the justification of a perturbation method explained above. We first note that if we regard r as a constant, the second term of (13) is by (18), Sec. 2.1, the square of the angular frequency:

$$\omega^2 = \frac{|K|}{mr^3} \tag{14}$$

and if we factor out the same quantity in the third term, (13) simplifies to

$$\ddot{r} + \omega^2(r - 2\kappa) = 0 \tag{15}$$

This is the equation of sinusoidal oscillation by the variable r about the constant, unperturbed value 2κ, from which it is obvious that

$$2\kappa = r_0 \tag{16}$$

and the solution of (15) is

$$r - r_0 = A\cos\omega t + B\sin\omega t \tag{17}$$

The constants of integration are determined from the initial conditions $r(0)$ and $\dot{r}(0)$. The former is a matter of convenience, and we define the time origin at a point where the path $r(t)$ crosses the unperturbed circle r_0, which makes $A=0$; the latter is found from (8) as follows:

$$\dot{r}_0 = \omega B = \frac{vF_\theta}{F_r} = \frac{vq\dot{\psi}}{qE_c} = \frac{v\dot{v}\phi r}{c^2\phi} = \frac{v^2\omega r}{c^2} \tag{18}$$

where the first equality follows by differentiating (18), the next from (8), the next substitutes the Faraday and Colulomb forces, the next expresses the two forces in terms of the Coulomb potential [using (21), Sec. 2.3, for the Faraday field], and the final equality differentiates v by multiplying it by ω.

Hence $B=\beta^2 r$ [and, incidentally, γ in (1) equals β^2], so that

$$r = r_0\left(1+\beta^2\sin\omega t\right) \tag{19}$$

That is, the electron very slightly wiggles round the unperturbed classical circle $r=r_0$. (In the following we denote the unperturbed IAAD values by the subscript 0, but omit it in the case of β, since the difference between β and β_0 is of order β^2.)

The solution (19) looks like a simple sine function, and it is, if we approximate ω by ω_0 and β by β_0. However, in a less crude approximation it is a rather nasty implicit function reminiscent of the exact solution (4), for ω is a function of r even by the "unperturbed" approximation (14), and so is β through (20), Sec. 2.1.

We would therefore expect the solution of (12) to be given not only by (19) with $\omega=\omega_0$, but also by the higher harmonics of that frequency, i.e., (19) with ω equal to an integral multiple of (14).

There is no need to rely on gut feelings for this. Complicated as the implicit function (19) may be in general, it is simple and exact for the points $r=r_0$, where the trajectory of the electron crosses the IAAD circle. Solving (19) for these points of a positive crossing (with $\dot{r}>0$), we find n such points spaced at angles $\omega t=2\pi/n$ round the circle by setting $\omega=n\omega_0$, where ω_0 is given by (14) with $r=r_0$. Thus higher harmonics are also a solution, so that the possible electron paths are given by

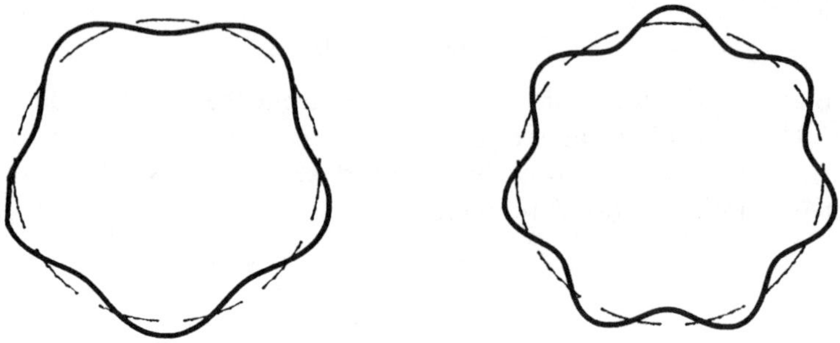

Electron orbits given by (19) for $n = 5$ and $n = 7$ with grossly exaggerated β.

$$r = r_0 \left(1 + \frac{\beta^2}{n} \sin n\omega t\right) \tag{20}$$

where the factor β^2/n follows again from the initial condition (8) as applied in (18) in determining the constant B in (17). This is shown in the figure with the amplitude of the wiggles drastically exaggerated. (In reality it is to the radius of the IAAD circle roughly as the height of an ocean liner is to the radius of the earth.)

The velocity of the electron in the transversal or (θ) direction is found from the total energy. We have

$$E = \frac{K}{2r_0} + \Delta E = -\tfrac{1}{2} m{v_0}^2 + \Delta E \tag{21}$$

where the energy additional to the unperturbed case

$$\Delta E = \tfrac{1}{2} m(v - v_0)^2 + \tfrac{1}{2} m\dot{r}^2 \tag{22}$$

If v_x is one of the velocities at the extremes of the trajectory (where $\dot{r} = 0$), then we have

$$\Delta E = \tfrac{1}{2} m(v_x - v_0)^2 = \tfrac{1}{2} m\dot{r}_0^2 = \tfrac{1}{2} m v_0^2 \beta^4 \tag{23}$$

where the second expression represents (22) on the IAAD circle, and the last expression follows from (9); comparing these two expressions we have

$$v_x = v_0(1 \pm \beta^2) \tag{24}$$

and since the velocity must be in antiphase with r,

$$v = v_0(1 - \beta^2 \sin \omega t) \tag{25}$$

Except for the fact that we now know the values of the constants A and B, the path (19) and the velocity (25) found here by dynamics are identical to the results found previously by electromagnetics.

However, beyond confirming the previous result, we have found an important additional point: the frequency ω, in this case the wiggle frequency, is no longer "some" frequency of electro-mechanical oscillations, but by (14) *it is equal to the electron's orbital frequency or an integral multiple of it.*

There is another way of expressing this. The length of the wiggles is nothing but the distance λ found in Sec. 2.2 from the geometric relation $v_0 = \lambda\nu$ which must be valid for any periodically varying velocity with mean v_0 and frequency ν, such as the special case (25). In the present case we have

$$\lambda = \frac{v_0}{\nu} = \frac{2\pi v_0}{\omega} = \frac{2\pi v_0}{n\omega_0} = \frac{2\pi r_0}{n} \tag{26}$$

that is, an electron orbit *must have an integral number of wiggles per orbit.*

From this, Bohr's postulates and the quantization of electron orbits follow quickly, as will be shown in the next section.

2.5 The Quantization of Electron Orbits

Let us define a stable orbit as one that passes through all points of the preceding one, or more briefly, an orbit for which $r(\theta)$ is a strictly periodic function with period 2π. This allows only closed orbits, not spiraling ones where the electron gains or loses energy.

However, it also excludes orbits which have a circle as their unperturbed "base," but whose undulations do not close on themselves (see figure), that is, orbits for which the number of wiggles is not an integer. (By definition of a period, a function with period $2\pi/n$, where n is an integer, automatically also has a period of 2π.)

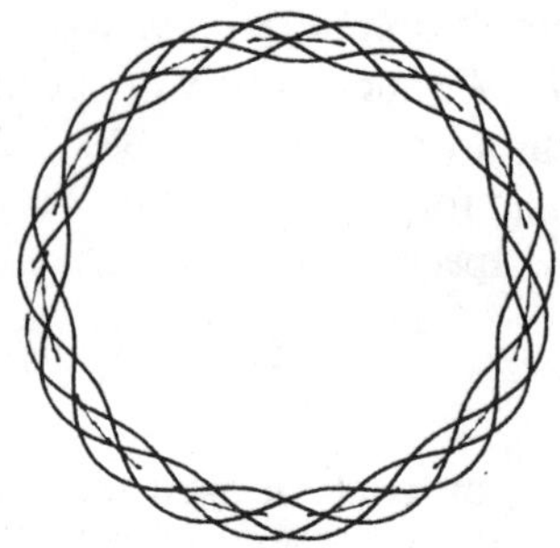

A non-integral number of wiggles per orbit, shown above, is incompatible with both the natural wiggle frequency and the initial condition (2), which follows form (22), Sec. 2.4

Such orbits are not just excluded by my definition of stability, but by the initial conditions under which our solution

$$r = r_0(1 + \beta^2 \sin \omega t) \tag{1}$$

is valid. It was obtained in the preceding section by satisfying the initial conditions

$$r(2n\pi) = r_0, \qquad \dot{r}(2n\pi) = \beta v \tag{2}$$

where the $2n\pi$ rather than 0 as in the previous section takes account of the electron going round more than once. Since the length of a full wiggle is by definition λ, the first of these conditions requires the length of the circumference $2\pi r$ to be an integral multiple of $\lambda/2$, and the second (always positive) disqualifies the half-wiggles from this set, so that we must have

$$2\pi r = n\lambda \tag{3}$$

where n is an integer. These two initial conditions thus limit the solutions, representing stable orbits, to those which contain an integral number of undulations.

The condition that the orbit must contain an integral number of "wavelengths" (wiggles), derived in the preceding section, is not taken from de Broglie; it is also found in violin strings, window panes, organ pipes, and electromagnetic resonators. In all of these, the natural frequencies – that is, the fundamental and its harmonics – are always those for which the geometric dimensions of the system contain an integral number of wavelengths. This behavior is dictated by the boundary conditions, which force a node of the waves to coincide with the geometric end of the oscillating source: mechanically by preventing a violin string or window pane from oscillating at its ends, or electromagnetically in a resonator by shortening the tangential component of the electric fieldstrength. The case of an electron orbit is similar: the integral multiple of oscillations per orbit is dictated by the initial conditions (2) in much the same way as in other vibrating systems.

In Newtonian IAAD mechanics, an orbit with *any* semiaxis is possible because it is *always* matched by a corresponding velocity given in general by (12), Sec. 2.1; for the case of a circle this reduces to

$$v^2 = \frac{K}{mr} \tag{5}$$

But the oscillations of a moving electric charge tie its velocity to the wiggle-wavelength λ by (8), Sec. 2.2, or

$$v = \frac{h}{m\lambda} \tag{6}$$

Thus, if the orbit is to be stable, the velocity now has *two* conditions to satisfy, namely (5) and (6), where λ must satisfy (3). The first, (5), is purely mechanical and applies to *any* body in central motion, charged or not; the other, condition (6), is the result of oscillations due to the self-induced Faraday field. When both conditions must be satisfied, only certain discrete velocities (and the resulting radii, frequencies, energies and angular momenta) are possible: eliminating λ from (3) and (6), and then r from the result and (5), the possible velocities are

$$v_n = \frac{v_1}{n}, \qquad n = 1, 2, \ldots \tag{7}$$

with

$$v_1 = \frac{2\pi|K|}{h} \tag{8}$$

and this is Einstein Plus One, for the rest is plain and pleasant sailing through old-fashioned mechanics: substituting (7) and (8) in (5) yields the stable radii

$$r_n = r_1 n^2, \qquad r_1 = \frac{h^2}{4\pi^2|K|m} \tag{9}$$

The angular frequency is found from $\omega = v/r$:

$$\omega_n = \frac{\omega_1}{n^3}, \qquad \omega_1 = \frac{8\pi^3 mK^2}{h^3} \tag{10}$$

and the energy of the electron is from (11), Sec.2.1,

$$E_n = \frac{E_1}{n^2}, \qquad E_1 = \frac{2\pi^2 K^2 m}{h^2} \tag{11}$$

making the difference in energy levels in the transition of an electron from level k

to level n

$$\Delta E(k,n) = \frac{2\pi^2 K^2 m}{h^2}\left(\frac{1}{k^2} - \frac{1}{n^2}\right) \tag{12}$$

and finally, Bohr's *postulate* on the angular momentum $L = mvr$ follows from (7) through (9),

$$2\pi L_n = nh \tag{13}$$

These results, in which postulates are replaced by derivations, are rooted in a single phenomenon: the validity of the Maxwell equations for *field-referred* velocities, from which the electron oscillations follow. As will be shown in Sec. 2.11, these oscillations contradict the Einstein theory. (The modifications of charge density, mass, force and energy, as derived from the Maxwell equations in Secs. 1.5 to 1.8 play no part here, and the meticulous reader may let m, for example, stand for the dynamic rather than the rest mass).

As for the results just obtained, Niels Bohr [1913] was right in everything but the clause "While there obviously can be no question of a mechanical foundation of the calculations given in this paper. . ."

Note that among all of the Bohr equations above there is only one reliable door into the atom, namely equation (12). When the left side is replaced by $h\nu_r = hc/\Lambda$, where Λ is the wavelength of the radiated light (do not confuse with the wiggle-length λ), we have the relation that can be checked against what is observed in a spectroscope:

$$(k-n)h\nu_r = \frac{4\pi^2 K^2 m}{h^2}\left(\frac{1}{k^2} - \frac{1}{n^2}\right) \tag{14}$$

This relation shows excellent agreement with the observed spectral series, especially when Bohr's theory is refined by using the reduced mass of the electron (rotating not about the nucleus, but about the common center of mass of the two – a refinement Bohr did not use, and neither was it used in Sec. 2.1, as it introduces needless complications).

Otherwise there are only some rough estimates, such as the order of r deduced from the scattering of light in gases and from the kinetic theory of gases. Everything else is *inferred* from (14). There is, for example, no direct way of measuring the orbital frequency

$$f = \frac{\omega}{2\pi} = \frac{v}{2\pi r} \tag{15}$$

We can find it from (14) by substituting (7) and (9) in the last expression (5) and comparing it with $\nu = \Delta E/h$. For a transition to an adjacent orbit, this yields

$$\nu_r = \pi f \frac{2n+1}{(n+1)^2} \tag{16}$$

On the other hand, we know from our derivation that the wiggle frequency ν must be an integral multiple of the orbital frequency f [this also follows immediately from (3) by setting $\lambda = v\nu$], so that

$$v = fn \tag{17}$$

In the Bohr theory, there are two distinct frequencies: the orbital frequency f, and the radiated frequency ν_r. In the present theory there is a third, the wiggle frequency ν. The relation among them is given by (15) to (17). The radiated frequency in (14), which is determined by the difference in energies, i.e., by the difference in orbital frequencies, is the only one of the three frequencies that can be directly found from the measured wavelength. The other two are inferred.

The question of why an electron does not radiate when it is in stable orbit will be discussed in Sec. 2.10.

2.6. Electromagnetic Mass

Inertia is the property of resisting acceleration: an impressed force **F** evokes an inertial reaction in the opposite direction.

However, as mentioned in Sec. 1.7, there are two types of inertial reaction to the force that, say, throws (accelerates) a tennis ball. If the tennis ball is uncharged, the only reaction is $m_n \dot{\mathbf{v}}$, where m_n is the Newtonian inertial mass of the ball and $\mathbf{v}$ its velocity. But if the ball is charged, the Faraday force (Faraday field times charge) will cause an *additional* inertial reaction, for it is proportional to the acceleration just as the Newtonian inertial reaction of uncharged matter is. *Additional* work must be done to accelerate a charged tennis ball to the same velocity as an uncharged one, the additional energy appearing in the magnetic field of the moving ball. The fact that in the case of a tennis ball the additional reaction and work is numerically negligible (as we shall see below) changes nothing in the principle, and in the case of an electron these additional quantities become significant.

In Sec. 1.5 we established that the electromagnetic field of a moving charge has a momentum

$$m_f = \frac{\epsilon}{c^2} \int (E_y^2 + E_z^2)\, dV \tag{1}$$

where m_f is the electromagnetic mass or field mass given by

$$m_f = \frac{\mu q^2}{6\pi R} \tag{2}$$

We derived this expression in Sec. 1.5 only for slow velocities, and in Sec. 1.7 we obtained a more general expression, namely (2) divided by $\sqrt{(1-\beta^2)}$, in agreement with the Einsteinian expression for mass. However, from now on we will mostly be concerned with velocities no higher than that of an electron in orbit at ground level ($\beta = 0.007$), and we will use (2) without the square root.

We can also derive the field mass of a charged sphere as in Sec. 2.2, by calculating the energy stored in the magnetic field when the sphere moves with velocity $\mathbf{v}$. The energy of the magnetic field is

$$U = \iiint \tfrac{1}{2} \mu H^2\, dV \tag{3}$$

where the volume V is all space outside the sphere with radius R, which is assumed to carry a surface charge q.

On substituting

$$H = \frac{E_c v \sin\theta}{\mu c^2}, \qquad dV = 2\pi r^2 \sin\theta\, dr\, d\theta \tag{4}$$

where E_c is the Coulomb field (Sec. 2.3) and integrating over θ from 0 to π and over r from R to ∞, we find

$$U = \frac{\mu q^2 v^2}{12\pi R} = \tfrac{1}{2} m_f v^2 \tag{5}$$

whence

$$m_f = \frac{\mu q^2}{6\pi R} \tag{6}$$

in agreement with (2).

However, both methods assume a constant velocity, and to make triply sure that we have the correct formula, we will now calculate the field energy of a moving sphere (5) by considering the work done when the charge is *accelerated* from velocity zero to velocity $\mathbf{v}$ against the resistance of the Faraday field.

Since this work appears as the energy of the magnetic field, we would expect to obtain agreement with (5), but direct verification is tricky, for we would have to integrate over the delayed $\boldsymbol{\psi}$ field moving outward as the charge accelerates (which will actually be done in Sec. 2.7). However, there is at least one other way of verifying (6) without this complication, and that is to calculate the flow of the Poynting vector $\boldsymbol{\psi} \times \mathbf{H}$ across the surface S of a surface-charged, accelerating sphere into the space outside it. This involves only the field immediately adjacent to the charge, where no delay effects can yet be important.

If we calculate this flow for the time interval during which the velocity has increased from 0 to v, this will yield the energy expended in accelerating the velocity to that value.

The scalar value of $\mathbf{H}$ is given by (4); its direction is

$$\mathbf{v}_o \times \mathbf{r}_o = \boldsymbol{\varphi}_o \tag{7}$$

where φ is the "longitude" of the sphere having $\mathbf{v}$ as the north-south axis. The Faraday field is, from Sec. 2.3, given by

$$\boldsymbol{\psi} = -\frac{1}{c^2}\frac{d(\phi \mathbf{v})}{dt} = -\frac{1}{c^2}[\phi \dot{\mathbf{v}} + \dot{\phi}\mathbf{v}] \tag{8}$$

where, in general, the acceleration and velocity have different directions, though either is constant with respect to the $\theta = 0$ direction as the point on the sphere r, θ varies. We consider the case when the acceleration has the direction of the velocity, both being directed along the $\theta = 0$ axis.

We first note that the second term in (8) contains the factor

$$\dot{\phi} = \frac{d\phi}{dr}\frac{dr}{dt} = -\frac{\phi}{r} v \cos\theta \tag{9}$$

and when this is multiplied by an element of surface

$$dS = 2\pi R^2 \sin\theta \, d\theta \tag{10}$$

and integrated over θ from 0 to π, it will cause the second term in (8) to vanish in the integration of the Poynting vector about to be performed,

$$\iint (\boldsymbol{\psi} \times \mathbf{H}) \cdot d\mathbf{S} \tag{11}$$

so that no more than the first term in (8) will survive the integration. Moreover, since $\mathbf{H}$ has the direction of the latitude rings (φ), and $\mathbf{S}$ that of $\mathbf{r}$, the only component of the acceleration that survives the triple product in the integrand is the θ component; hence the Faraday field on the surface is

$$\boldsymbol{\psi} = -\frac{\phi \dot{\mathbf{v}} \sin\theta}{c^2} + \cdots = -\frac{\mu q \dot{\mathbf{v}} \sin\theta}{4\pi\epsilon R} + \cdots \tag{12}$$

where the three dots stand for the terms that are about to vanish in the integration.

We now integrate the flux of the Poynting vector (11) over time from the moment when the velocity is zero to the moment when it has reached a velocity v, obtaining the total energy that has moved from the field into the sphere (or, on deceleration, in the opposite direction):

$$\int_{t(0)}^{t(v)} \iint_S (\boldsymbol{\psi} \times \mathbf{H}) \cdot \mathbf{H}\, dS\, dt = \frac{\mu q^2 (2\pi R)^2}{(4\pi)^2 R^3} \int_0^v \int_0^\pi v \sin^3\theta dv\, d\theta = \tfrac{1}{2} m_f v^2 \tag{13}$$

where m_f is once again given by (6). This shows, as intended by this little side check, that (6) holds not only for a uniformly moving charge, but also for an accelerated charge involving the Faraday field (via the Poynting vector).

Let it, however, be stated straight off that the flow of a Poynting vector across the surface S does not imply radiation in this case. It is immediately evident from (12) and (4) that $\boldsymbol{\psi}$ and $\mathbf{H}$ are proportional to $\dot{\mathbf{v}}$ and $\mathbf{v}$, respectively, so that in our case, when $\mathbf{v}$ oscillates in time, the two are 90° out of phase, causing the Poynting vector to change direction twice per cycle. This means, exactly as it should, that energy is flowing back and forth between field and electron as the energy of the field changes to kinetic energy and back.

Such an oscillatory Poynting vector is very different from the case of $\mathbf{E}$ and $\mathbf{H}$ in a propagating electromagnetic wave (or from voltage and current in the radiation resistance of an antenna in which the electrons undergo *forced* oscillations by externally supplied energy). In that case the two are in phase, so that both change direction simultaneously, leaving the direction of the Poynting vector and hence of the energy flow unchanged, namely outward in the direction of propagation.

Let us now return to the electromagnetic or Faraday mass (6) and apply it first to a tennis ball. If it has a radius of 5 cm (any accuracy in this example would be entirely misplaced), and can withstand a voltage of 10 kV, then it can be given a charge of 0.06 microcoulombs, which by (6) will give it an electromagnetic mass of 4×10^{-21} kg. If it weighs 100 g, that is roughly 10^{-20} of its Newtonian mass, and therefore utterly insignificant.

In the case of an electron, which by (2) does have a significant Faraday mass, things would be equally simple if we were able to charge and discharge an electron like a tennis ball. Thus the experimental resolution of the two masses is not feasible for macroscopic matter because the Faraday mass is too small in comparison with the Newtonian mass; and elementary particles cannot be discharged (at least not while leaving all their other properties unaltered).

Nevertheless, the example of the tennis ball shows that – if the Maxwell equations are correct – the resolution of the two types of mass is not *inherently* impossible: the inability of measuring with an accuracy of one part in 10^{20} is due to technical imperfection, not due to the constraints set by a natural law.

In the following we shall attempt such a resolution, and as will be shown (more carefully than in Sec. 2.2), this can be achieved on theoretical grounds.

We must first go back to basics.

An *impressed* force (that which changes the momentum of a body) such as the muscle force throwing an *uncharged* tennis ball, provokes an inertial reaction in the body on which it is acting. Since action and reaction are equal and opposite, this yields the familiar equation

$$\mathbf{F} = m_n \mathbf{a} \tag{14}$$

where **a** is acceleration. (Here and in the following second-order modifications are omitted as an unnecessarily accurate distraction that sheds no light on the issue to be examined.)

The two sides of (14) are numerically equal, as they must be in any equation, but conceptually they are different: the left side is the impressed force (that which causes a body to change its state of rest or uniform motion), the right side is the reaction to it. The m_n in (20) stands for the Newtonian mass, that is, the resistance to acceleration presented by the body when it is not charged.

When the body on which the force acts is charged and free to accelerate (that is, when the sum of impressed forces does not add up to zero as it does in the static case), then there appears an additional Faraday force, which is opposed by its own inertial reaction:

$$q\boldsymbol{\psi} = m_f \mathbf{a} \tag{15}$$

where m_f is the Faraday or field mass given by (2).

This equation may require some additional explanation. When the force on a charged body is the Coulomb force, which is independent of (small) velocity, or the friction of the air, which is proportional to the square of the velocity, then (14) is clear enough. But when the force is proportional to the acceleration, the resulting equation $m_f a = m_f a$ is seemingly trivial and can be confusing, although the left side, just as in (14), is the applied force, and the right side, again as in (14), is the inertial reaction to it. The confusion may arise through uncertainty whether the Faraday force is merely an increased inertial reaction (causing an increased mass), or whether it is a genuine additional *impressed* force with its own additional inertial reaction.

If the Lorentz force is valid, then $q\boldsymbol{\psi}$ is, of course, an impressed force just as $q\mathbf{E}_c$ is; but the question is also answered by the sign of the force: if it has the same sign (direction) as the acceleration, then it is another "impressed" force; if it opposes the acceleration, then it is an inertial reaction (which by our temporary convention belongs on the other side of the equation). The Faraday force, by (21), Sec. 2.3, is

$$q\boldsymbol{\psi} = -\frac{q\dot{\phi}}{c^2} \tag{16}$$

where ϕ is the Coulomb potential. Due to the acceleration and the delays introduced by it, this potential differs slightly from the electrostatic potential, but not enough to change the sign, which remains negative, so that (16) has the same direction as the acceleration. Hence it is an *impressed* force, not a reaction on the "wrong" side of the equation. The reaction to that force, which has the same sign (on the other side of the equation), is then given by the right side of (15).

Here I will insert a little "hackle smoother." By Lenz's Law, the direction of the induced electric field *opposes* its cause, which in the case of a tennis ball is the acceleration imparted to it by the thrower. Yet we have just found that the Faraday force $q\boldsymbol{\psi}$ acts *in the direction* of the acceleration. Is that a contradiction?

No. What opposes the acceleration, i.e., the applied force, is the electromagnetic inertial reaction $m_f\mathbf{a}$; the Faraday force is the *additional* force required by the thrower's muscles to overcome this reaction of the field. If the Faraday force went against the acceleration, both that force and the inertial reaction to it would subtract from the original force and its inertial reaction and make the throwing of a charged tennis ball easier, not harder. The only thing wrong here is the fuzzy or incorrect formulation of Lenz's Law.

Now consider the seemingly innocent equation

$$q\mathbf{E} = m\mathbf{a} \tag{17}$$

which blooms out into

$$q(\mathbf{E}_c + \boldsymbol{\psi}) = (m_n + m_f)\mathbf{a} \tag{18}$$

The significance and misinterpretation of this relation is perhaps best discussed by means of an analogy which has nothing to do with either Einstein or electromagnetism.

Let an uncharged tennis ball with mass m be shot out of a slingshot with force kx (spring constant of the rubber k, expansion x), so that in analogy with (14) we have

$$kx = ma \tag{19}$$

Now imagine that inside the ball there is a little rocket engine controlled by a computer that senses the external acceleration and adjusts the thrust to be directly proportional to it, developing an additional forward force

$$P = Ma \tag{20}$$

Here M is a constant of proportionality, which clearly has the dimension of mass, but is not actually equal to the Newtonian mass of the projectile (for its value is obviously in the hands of the engine designer and computer programer). M, then, is not a quantity of matter, but a virtual mass playing the same role as a genuine mass in determining the resistance to acceleration. Relation (20) is the analog of (15) and (16).

There is only one common acceleration and one common velocity for the projectile consisting of tennis ball and engine, so that the equation of motion is

$$kx + P = (m + M)a \tag{21}$$

which is clearly the analog of (18). The work done by the elastic rubber and the engine to achieve a velocity v as the projectile leaves the slingshot (the "muzzle velocity" if it were a gun) is found by integrating (21) over the stretch distance x:

$$\int kx\,dx + \int P\,dx = \int (m + M)\,ma\,dx \tag{22}$$

or

$$W_1 + W_2 = \tfrac{1}{2}(m + M)\,v^2 \tag{23}$$

So far, I hope, mechanical engineers will agree. But now comes a nasty (yet, as we shall see, very pertinent) question: What happens to the man who wishes to determine the mass of the projectile by measuring the muzzle velocity v and the energy W_1, but who ignores the energy W_2?

He will infer an incomplete mass m_i following from the incomplete energy on the left, namely

$$m_i = \frac{2W_1}{v^2} \tag{24}$$

But this is exactly the value he would find by integrating (19); that is, the mass given by (24) is the Newtonian inertial mass only. As long as the man is aware of this, there is no harm done, for no inconsistency threatens. But if he believes that (24) represents the ratio of the entire force to acceleration, or the mass as the usual coefficient of the entire kinetic energy on the right of (23), he will be in for some surprises. For example, if he fires the projectile into a calorimeter, where its entire energy is dissipated into measurable heat, he will find the energy in excess of the expected kinetic energy imparted by the slingshot and excluding that imparted by the little rocket engine.

If he is a modern physicist, he will claim to have found a phenomenon that classical physics cannot explain, and conclude that tennis balls fired from sling-shots have their own physics.

The way the measurements of electron mass are interpreted are disturbingly close to my little story.

There are several methods of measuring the electron mass, and within the experimental error they all yield the same result: the unspecified "electron mass" equals 9.81×10^{-31} kg. Modern methods include electron scattering by a grating; the older methods are based on observing electrons accelerated by the Lorentz force in an electric or magnetic field. The more usual method uses a magnetic field and finds the mass from the electron charge and from the ("cyclotron") frequency with which the electrons will circle. The method that is closest to my little story is an almost forgotten, but witty setup used by Kirchner [1931], which measures the mass of the electron using only an electric field.

Kirchner, whose experiment is shown in the figure and explained in its caption, finds the electron "mass" m by equating the kinetic and potential energies as

$$m = \frac{2qV}{v^2} \tag{25}$$

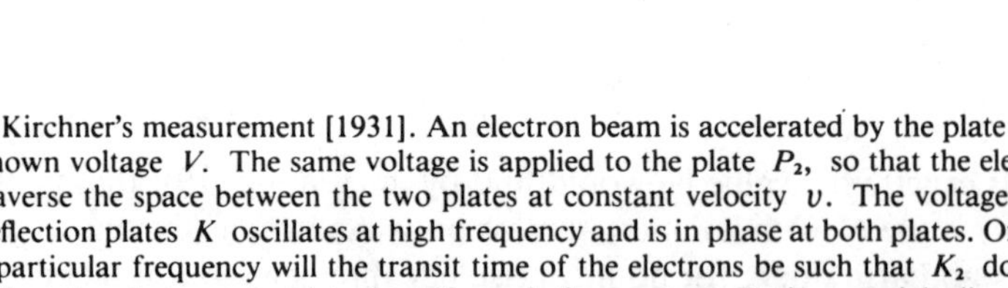

Kirchner's measurement [1931]. An electron beam is accelerated by the plate P_1 at known voltage V. The same voltage is applied to the plate P_2, so that the electrons traverse the space between the two plates at constant velocity v. The voltage at the deflection plates K oscillates at high frequency and is in phase at both plates. Only for a particular frequency will the transit time of the electrons be such that K_2 does not deflect the electrons, but lets them through the aperture D_2 in a straight line to the screen S; for nearby frequencies, the beam will be deflected from this point. When this frequency and hence the transit time have been established, the known distance determines the velocity of the electrons, and their mass is then found by equating their kinetic energy to the potential energy at the plates P. Unlike measurement in a magnetic field, which accelerates the electrons transversally, Kirchner's method measures their mass under longitudinal acceleration. Both methods, as do all others, yield the same result.

where q is the electron charge and V the voltage of the accelerating plates P. This interpretation of the electron mass is thus precisely equal to (24): *This and all other measurements of "electron mass" measure only the Newtonian mass of the electron*, that is, the inertial mass it would have if it were not charged. It takes no account of the fact that, if the Maxwell equations are valid, the electron must also have an electromagnetic mass: if Kirchner's electron beam had ended up in a calorimeter, it would have measured more than the energy (25).

This electromagnetic or Faraday mass is given by (6), but just as in the case of a charged tennis ball it proved to be too small to be experimentally observable, so in the case of the electron it is not usable, because it involves the electron radius, a controversial quantity, which, with circular irony, depends on the electron mass.

What we can do, however, is imitate our hypothetical physicist and measure the total energy of a moving electron; for the reasons explained above, this energy should be too large. But instead of blaming the discrepancy on classical physics, we will use it to infer the Faraday mass in (18), or closer to the analogy, M in (23).

We start from the kinematic "carrot formula" (1), Sec. 2.2, for the electron moving with oscillating velocity

$$v = \lambda\nu \tag{26}$$

On the false premise that the entire energy of the moving electron is its mechanical (Newtonian) kinetic energy, and that what we measure in Kirchner-type experiments is the entire mass of the electron, we have

$$W = [??]\ \tfrac{1}{2}m_n v^2 = \tfrac{1}{2}m_n \lambda v\nu = \tfrac{1}{2}h\nu \qquad \text{[false]} \tag{27}$$

where

$$h = m_n \lambda v \tag{28}$$

is "a" constant, which in Sec. 2.2 emerged simply as a constant of proportionality expressing the conservation of energy. In fact, of course, it is Planck's constant, as shown by its occurrence in the Bohr orbit formulas, and as also shown by the numerous successful applications of (28), which is identical with the well tested de Broglie hypothesis.

But when (27) is compared with the experimentally measured values, it is clearly in error: the measured energy of an electron is twice as large. From this we must conclude that *the total mass of an electron is twice as large as its Newtonian mass, i.e., the Newtonian and Faraday masses of an electron are equal:*

$$m = m_n + m_f = 2m_n = 2m_f \tag{29}$$

Contemporary orthodoxy rarely asks how an electron's total inertia is divided into Newtonian and electromagnetic; it accepts de Broglie's successful hunch as one of the laws of nature rooted in experiment (or alternatively points to the resulting Schrödinger Equation, which is another "that's the way it is" relation). There is nothing wrong with that approach in itself, for all natural laws must ultimately be rooted in experience; what is wrong is the failure to ask why a charged tennis ball does not obey (28). The conventional reply invokes the Correspondence Principle, by which the laws governing the particles of the microworld merge with classical laws as large numbers of such particles are subjected to the corresponding statistics. But a very large number of electrons in a beam still obeys (28) without converging to any classical law, so we still have one physics for electrons and another for charged tennis balls.

By contrast, the present theory provides a simple answer: a charged tennis ball would indeed obey (28) if its ratio q^2/R could be increased to a value sufficiently large to make its electromagnetic inertia significant in comparison with its mechanical inertia. An electron, on the other hand, *always* has that ratio sufficiently large: its electromagnetic inertia is 50% of the total.

There is only one physics.

2.7. Electromagnetic mass and acceleration

The expression for electromagnetic or field mass of a spherical charge q with radius R

$$m_f = \frac{\mu q^2}{6\pi R} \tag{1}$$

was found in three different ways: from the electromagnetic momentum (Sec. 1.5), from the energy of the magnetic field surrounding a moving charge, and from the Poynting vector of a charge accelerated from zero to velocity v (Sec. 2.6).

We shall now derive (1) by a very simple method from first principles: by calculating the force on an accelerated charge by its own electric field.

Consider a spherical surface charge with radius R at rest or in uniform motion. The net force exerted by the resulting Coulomb field on its source charge is clearly zero: the forces in any direction cancel. As a model for what is to follow, we will go through this trivial calculation: the force in, say, the x-direction is

$$F_x = \int E_x \rho \, dS \tag{2}$$

where E is the electric field at the surface of the sphere, ϱ is the surface charge density $(q/4\pi\epsilon R^2)$, and dS is an element of surface $(2\pi r^2 \sin\theta)$. Hence

$$F_x = \tfrac{1}{2} E_0 \int_0^{\pi} \sin\theta \cos\theta \, d\theta = 0 \tag{3}$$

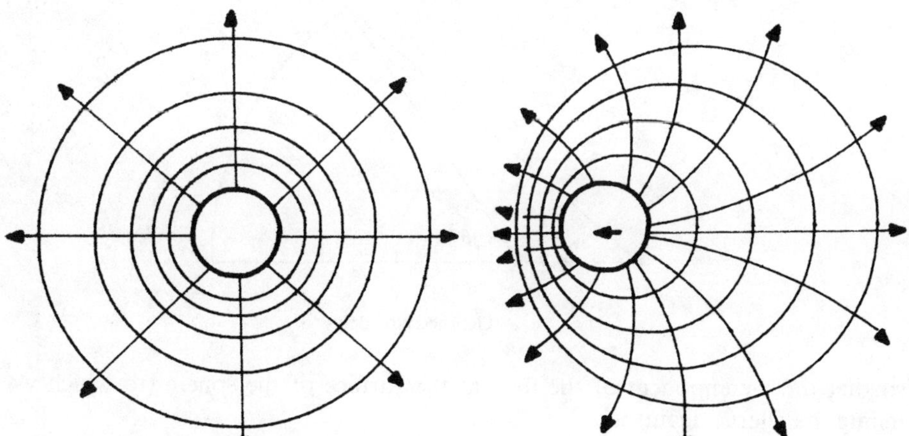

Equipotentials and electric field of a uniformly moving and of an accelerated charge

Now let the charge be accelerated with constant acceleration (and for sufficiently small time intervals we can consider *all* accelerations constant) in the x-direction, as shown in the right part of the figure. As the charge accelerates, it will partly catch up with its own equipotential spheres, propagating from the charge with velocity c; these spheres are retarded in the sense that their centers are at the points where the charge was at the time of their emission. The electric field lines, being orthogonal to this family of spheres, are now curved — the total electric field equals the sum of the Coulomb field as on the left, plus the Faraday field evoked by acceleration.

Clearly the net force on the source charge no longer cancels in all directions: the component in the direction perpendicular to the acceleration still adds up to zero, but in the direction of the acceleration, the force is asymmetrical, since the gradient of the potential is larger on the leading side of the moving charge than on the trailing one.

To calculate the net force, we again use (2), but this time E is no longer the pure Coulomb field. Using the Divergence Theorem to shrink the charge to a point, we note that the radius of an equipotential sphere r is no longer the distance from a point on that sphere to the charge, for during the time the potential has propagated to the distance R, the equivalent point charge has moved forward by some small distance s. The distance effective for the electric field is therefore

$$d = \sqrt{R^2 + 2Rs\cos\theta + s^2} \tag{4}$$

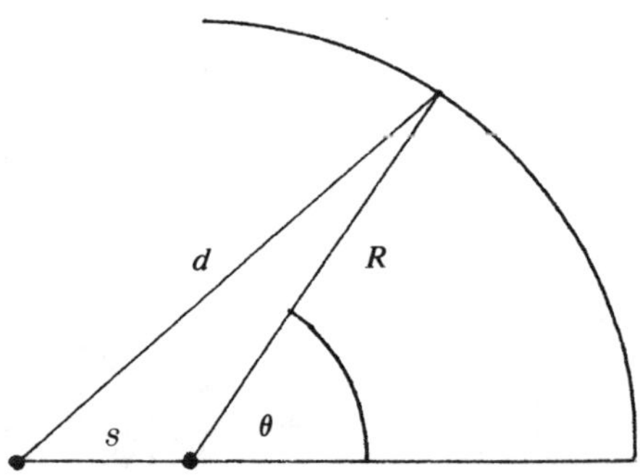

Geometry of delay

so that the x-component of the field at the surface of the sphere (to which we inflate the charge again) is

$$E_x = \frac{q\cos\theta}{4\pi\epsilon_0 R^2[1 + 2(s/R)\cos\theta + (s/R)^2]} \tag{5}$$

or, since $s<<R$, we have to first order

$$E_x \approx E_0 \left(1 - \frac{2s}{R}\cos\theta\right)\cos\theta \tag{6}$$

Now s is the distance to which the point charge moved from rest under constant acceleration during the time the retarded potential, advancing with the constant velocity of light, reached the distance R; hence

$$s = \tfrac{1}{2}\dot{v}t^2 = \tfrac{1}{2}\dot{v}\frac{R^2}{c^2} \tag{7}$$

Substituting this in (6) and the result in (7), we have, since the fiist term integrates to zero,

$$F_x = \int_0^{\pi} \frac{\mu q^2 \dot{v}\cos^2\theta\sin\theta}{4\pi R}\,d\theta \tag{8}$$

Since

$$\int_0^{\pi} \cos^2\theta\sin\theta = \tfrac{2}{3} \tag{9}$$

and there is no force other than that in the x-direction, we finally have

$$\mathbf{F} = \frac{\mu q^2}{6\pi R}\dot{\mathbf{v}} \tag{10}$$

in which the factor multiplying the acceleration is identical with (1).

Thus the field mass is independent of acceleration and identical to that found as the ratio of electromagnetic momentum to velocity, or double the magnetic energy to the square of velocity – in all cases to first order in β.

The field mass is therefore a full-fledged mass that appears in the expressions for force, momentum and energy in the same way as Newtonian mass – in the case of energy, it is called *magnetic* when associated with the field of a uniformly moving charged body, but *kinetic* when associated with Newtonian, uncharged matter.

It is noteworthy that the derivation of the mass in (10) is based on first principles, including the propagation of force and potential with velocity c. Otherwise it uses no more than the Lorentz force $\mathbf{F}=q\mathbf{E}$ (the magnetic field of the moving charge would introduce a second-order correction), the relation $c^{-2}=\sqrt{\epsilon\mu}$, and the equation $\operatorname{div}\mathbf{D}=\varrho$. If, for example, the other Maxwell equations were invalid, the derivation of (10) given here would still hold.

2.8. Energy Balance

Let us now check that the undulating electron orbit satisfies the energy balance. The path and velocity of an electron in orbit derived in Sec. 2.4,

$$r = r_0(1 + \beta^2 \sin \omega t) \tag{1}$$

$$v = v_0(1 - \beta^2 \sin \omega t) \tag{2}$$

would not satisfy the energy balance (i.e. keep the total energy constant) if we overlooked a rarely met quantity, the energy of the displacement current.

An electron (or other charge) in motion is usually thought of as equivalent to a conduction current, and indeed, this is how we found its magnetic energy: we determined its magnetic induction **B** from its own Coulomb field, and the integral of $\frac{1}{2}B^2/\mu$ over all space then yielded the energy

$$T_f = \tfrac{1}{2}mv^2, \qquad m_f = \frac{\mu q^2}{6\pi R} \tag{3}$$

This is exactly analogous to the mechanical kinetic energy of uncharged matter (such as the one carrying the charge), and quite similarly, the magnetic-kinetic (or "quasikinetic") energy of the radial component is $\frac{1}{2}\dot{r}^2$.

But the radial component also has another type of energy associated with it. It is reminiscent of the potential energy of a charge in the field of another charge and is related to the displacement current $\partial \mathbf{D}/\partial t$. As the electron changes its radial distance from the proton, the Coulomb field in its neighborhood changes; this constitutes a displacement current, which by the second Maxwell equation has a magnetic field associated with it, and its energy can then be calculated in the same way as we did with the quasikinetic energy. The resulting "quasipotential" energy is, like that of the quasikinetic energy, again proportional to the square of the radial velocity; however, while the quasikinetic energy is based on the magnetic field associated with the motion of the electron's *own* Coulomb field, the quasipotential energy is associated with the change of the Coulomb field the electron is traversing. Perhaps surprisingly (since it is based "only" on the displacement current), the quasipotential energy turns out to be of the same order as the quasikinetic energy.

With **E** the Coulomb field of the proton at a distance r from the latter, we have in the neighborhood of the electron

$$\frac{\partial \mathbf{E}}{dt} = \frac{\partial \mathbf{E}}{dr}\dot{\mathbf{r}} = -\frac{q}{2\pi r^3}\dot{\mathbf{r}} \tag{4}$$

Therefore the second Maxwell equation, simplified by the relation $c^2 = 1/\epsilon\mu$, becomes

$$\nabla \times \mathbf{B} = -\frac{q\mu}{2\pi r^3}\dot{\mathbf{r}} \tag{5}$$

The lines of **B** curl round the radial velocity, i.e., round the position vector **r**, and we can find the value of B near the electron from Stokes' Theorem by using the circles of **B** with radius $r\sin\theta$ as the closed path of integration:

$$2\pi r \sin\theta B = -\frac{q\mu\pi r^2 \sin^2\theta}{2\pi r^3}\dot{r} \tag{6}$$

or

$$B = -\frac{\mu q \sin\theta}{4\pi r^2} \tag{7}$$

The energy could now be found by integrating the magnetic energy $\tfrac{1}{2}B^2/\mu$ over all space outside the electron as we did in Sec. 2.2; but more simply, we note that (7) is, except for the negative sign, clearly equivalent to

$$\mathbf{B} = \frac{\mathbf{v}\times\mathbf{E}}{c^2} \tag{8}$$

which was our starting point in Sec.2.2. Thus the result must be the same, namely (3).

Since both the quasikinetic and quasipotential energies of the electron are due to its motion in the field of another charge, and both have the same value, the suspicion may arise that we have merely counted the same thing twice. That this is not so can quickly be demonstrated by considering the radial motion of an electron in a field that does not obey the inverse square law – for example, in the field of a dipole (proportional to $1/r^3$). Then the quasikinetic energy of the electron remains unchanged as given by (3), but clearly the derivation (4) through (7) will now lead to a different result, showing that the two energies are different.

There is, however, another point to be considered: the electron will do unto the proton as the proton does onto the electron – they will each give rise to displacement currents in the neighborhood of the other. (We recall that the only effect-producing velocities are those with respect to the local field; the observer has no standing.) But now we run into a problem: what should we substitute for R in (3) when the moving charged body is a proton rather than an electron?

One way of evading the difficulty would be to speculate that a proton is really a neutron plus a positron, and a positron presumably (more speculation) has the same radius as an electron.

But that glosses over a more important point, namely that not everything is in order with R even for an electron, if it is literally understood to be the radius of a ball carrying an electric charge on its surface. If this were so, then that radius, from (3), using the equality $m_f = m$, would be

$$R = 1.8787 \times 10^{-15}\ \mathrm{m} \tag{9}$$

but electron-electron scattering experiments show that the genuine electron radius must be much smaller. On the other hand, I believe that the reader will find the derivation of (3) in Sec. 2.2 as unimpeachable as the Maxwell equations.

To resolve the paradox, we note that the Divergence Theorem permits any charge with radial symmetry to be inflated or shrunk to any size radius without changing its external field.[1] The question therefore arises whether R is the genuine radius of a genuine charged sphere, or whether it is simply an equivalent, "just-as-if radius" that will produce an observed or calculated field.

But that question is easily answered: we are all convinced that a proton and electron – about whose geometric properties we know little or nothing – moving at the same velocity will, except for sign, produce the same magnetic field; indeed, it follows from the Divergence Theorem that two tennis balls with the same charge moving at the same velocity will produce the same magnetic field at the same external distance from their centers, regardless of their radii (provided only they are shorter than the considered distance, which is why it must be "external"). It therefore follows that R is not something that determines the field, but something that *is determined by* the field and has no simple relation to the actual and precise geometric radius.

In particular, since proton and electron have the same charge, and for the same velocity produce the same absolute magnetic field, they must also have the same equivalent radius R.

Returning to (8), this means that the total quasipotential energy of the electron in orbit must be twice the magnetic energy corresponding to (8), or

$$U_f = -m_f \dot{r}^2 \tag{10}$$

The negative sign follows from the fact that (7) has the opposite sign from (8); the corresponding energy must therefore be subtracted from the quasikinetic energy, which we will take as positive.

With the energy of the displacement currents established, the energy balance is now easily checked. Denoting the total mass (Newtonian plus electromagnetic) by m, the kinetic-magnetic energy by T, the potential and quasipotential energies by U and U_f, and the total energy by E, we have

$$E = T + U + U_f = \tfrac{1}{2}mv^2 + \tfrac{1}{2}m\dot{r}^2 + \frac{K}{r} - m\dot{r}^2 \tag{11}$$

Using the classical IAAD relation (20), Sec. 2.1,

$$\frac{K}{r_0} = -mv_0^2 \tag{12}$$

[1] One can even play with the Divergence Theorem *inside* the charge distribution. For example, assuming the earth to have radially symmetrical density, it is easily shown that the gravitational field at the bottom of a deep mine equals the field that would result if the earth's mass below that depth were shrunk to a point at the center of the earth and the layer above it were removed.

substituting (1) and (2), and expanding U up to terms of fourth order in β, we then obtain

$$\begin{aligned} E &= -\tfrac{1}{2}mv_0^2[(1-\beta^2\sin\omega t)^2+\beta^4\cos^2\omega t] \\ &\quad - mv_0^2(1-\beta^2\sin\omega t+\beta^4\sin^2\omega t) - mv_0^2\beta^4\cos^2\omega t \\ &= -\tfrac{1}{2}mv_0^2(1+\beta^4) \end{aligned} \tag{13}$$

showing that the total energy is constant to terms of fifth order (for β^4 is, to that extent, constant over an orbit with circular base). It differs from the classical total energy by a factor of $(1+\beta^4)$.

Thus even if our solution (1) and (2) were wrong (and the next section will show that it is not), it would at least satisfy the energy balance, i.e., make the total energy constant throughout the orbit.

This result holds equally well for orbits with several wiggles per orbit, since the derivation remains unchanged if ω is replaced by $n\omega$, and β^2 by β^2/n; the energy for the higher orbits therefore also remains constant, but at the value

$$E = -\tfrac{1}{2}mv_0^2\left(1+\frac{\beta^4}{n^2}\right) \tag{14}$$

It is noteworthy that

1) neither the kinetic, nor potential, nor quasipotential energy remain constant separately – only the sum does;

2) the orbits with several wiggles have more energy mainly due to the classical reason that the velocity is lower and the radius larger – the wiggles themselves change the energy only by a fraction β^4/n of the zero-order energy.

3) the wiggles slightly *decrease* the IIAD energy $-\tfrac{1}{2}mv^2$.

The first of these conclusions confirms what was taken as obvious in Sec. 2.5: a stable orbit requires not just the classical (mechanical), but also the electromagnetic energies to be balanced.

2.9. Planck's Constant

Max Planck's discovery of the quantization of energy was based on black-body radiation; via Bohr's model of the atom, the de Broglie hypothesis, and the Schrödinger equation, this led to quantum mechanics, and Planck's constant $h = 6.6262 \times 10^{-34}$ Js is today recognized as one of the fundamental constants of nature.

But fundamental constants differ when classified by the level of our understanding. James Clerk Maxwell's discovery that $c = \sqrt{\epsilon\mu}$, for example, revealed light to be an electromagnetic phenomenon. On the other hand, the gravitational constant Γ is a "that's the way it is" number of as yet unknown origin. Planck's constant has been in the same category.

It will be shown below, first, that Planck's constant can be derived from Maxwell's equations without additional hypotheses, and second, that it is associated with the electric charge (the electron charge) of a particle, but is entirely independent of its mass or other properties.

It will be recalled that the constant h was introduced in the present theory without connection to black-body radiation, atomic spectra or quantum mechanics. It appeared in Sec. 2.2 as a lowly constant of proportionality in the energy equation

$$mv^2 = mv\lambda\nu = h\nu \tag{1}$$

which itself follows from the "carrot equation"

$$v = \lambda\nu \tag{2}$$

making

$$h = m\lambda v \tag{3}$$

Using Newton's Laws and Maxwell's equations, we then derived the Bohr orbits, in which this constant of proportionality appeared in all the places where Planck's constant stands when the orbits are derived from Bohr's postulates. The two constants were therefore recognized as identical.

But we can also aim directly for Planck's constant, bypassing the Bohr orbits, and using only the energy of an electron orbiting with velocity v_0 at ground level, i.e., with only one wiggle per orbit. This will provide a check on our previous results, as well as throw some light on the nature of Planck's constant.

The charge will oscillate, and its natural frequency of oscillation can be established from its energy as in any other vibrating system, quantized or not; Planck's constant then follows from (1).

We shall calculate the natural frequency from the energy that is being converted from the Faraday field to the magnetic field and back:

$$\int [\tfrac{1}{2}\epsilon\psi^2 + \tfrac{1}{2}\mu(H - H_0)^2]\, dV = \text{const} \tag{4}$$

or as we know from Sec. 2.2,

$$\int \tfrac{1}{2}\epsilon\psi^2\, dV + \tfrac{1}{2}m_f(v - v_0)^2 + \tfrac{1}{2}m_f(\dot{r} - \dot{r}_0)^2 = \text{const} \tag{5}$$

where

$$m_f = \frac{q^2\mu}{6\pi R} \tag{6}$$

Using (21), Sec. 2.3,

$$\psi = -\frac{q\phi}{c^2}\dot{v} \tag{7}$$

a straightforward integration yields

$$\int \tfrac{1}{2}\epsilon\phi^2 = \frac{3}{4}\frac{m_f R r}{c^2}\dot{v}^2 \tag{8}$$

where r is the as yet unspecified radius of the volume of a sphere forming the volume of integration about the accelerating electron. [Strictly speaking, we should also have included the square of the second term in (20), Sec. 2.3, but it turns out negligible compared with the second and third terms in (5).]

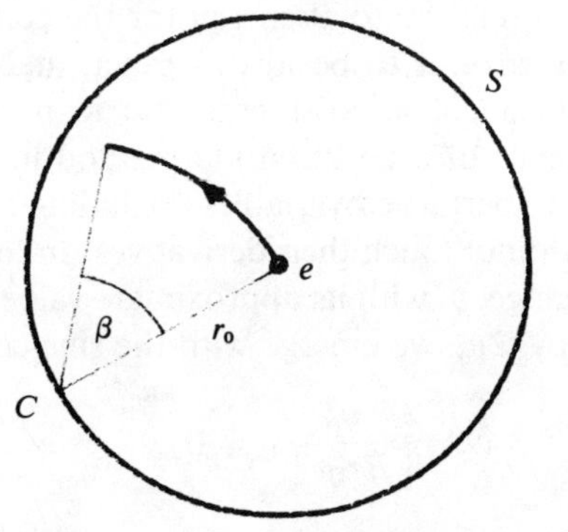

Geometry of energy balance calculation.

The electron is at e, the proton at C. It follows from elementary geometry that as the electron moves through an angle β, the potential (field, force) emanating from the electron with velocity c will have advanced to the surface of a sphere with radius r_0. The change in magnetic energy along the arc must therefore equal the energy of the Faraday field contained in the sphere S.

The integration (8) is "straightforward" only because it has been assumed that the acceleration $\dot{v}$ is constant and that the electron remains at the center of the sphere of integration, i.e., that there are no delay effects as in the figure on p. 114. Neither of these assumptions is true in general, but they are very close to being fulfilled when the electron path crosses the IAAD circle, e.g., shortly after time $t=0$. We therefore use an artifact based on elementary geometry: when a light source moves through a circular arc, its light reaches the center of the arc during the time it has moved through an angle β. Therefore the integral (8) will be a very close approximation if we set $r=r_0$; at the same time we must substitute the remaining two terms in (5) for $\omega t=\beta$. For small angles (not necessarily β) one finds

$$\tfrac{1}{2}m_f(v-v_0)^2+\tfrac{1}{2}m_f(\dot{r}-\dot{r}_0)^2=\tfrac{1}{2}m_f\beta^4v_0^2 \tag{9}$$

and when this, together with (6), is substituted in (5), we have the equation that will yield the natural frequency:

$$\frac{3}{4}\frac{\dot{v}^2Rr_0}{c^2}+\frac{1}{2}\frac{v_0^2v^4}{c^2}=\text{const} \tag{10}$$

It should be noted that all the symbols in this equation are constants; for example, v is the fixed value of the velocity after the electron has advanced from the IAAD circle through an angle β. However, by the rationale of a perturbation method, we may assume that the same relation holds *near* these fixed values, i.e., that in the neighborhood of these values (10) is not just a relation among constants, but a differential equation. The equation is that of Jacobian elliptic functions (sn and cn rather than sin and cos), from which the natural frequency could be determined on substituting the total energy for the constant on the right.

However, the result turns out to be just as exact, and the labor a good deal smaller, if we demote the Jacobian aristocrats to trigonometric peasants. This is again done by the rationale of a perturbation method in which, as explained in Sec. 2.4, perturbed and unperturbed variables (including constants) may be interchanged, as long as we do not touch their derivatives. In the present case, we may therefore simply interchange v with its approximate value v_0. On differentiating the result and dividing by $2\dot{v}$, we emerge with the sine-cosine equation

$$\ddot{v}+\frac{2}{3}\frac{v_0^4}{Rr_0c^2}v=0 \tag{11}$$

whence the required natural frequency is

$$\omega^2=\frac{2}{3}\frac{v_0^4}{Rr_0c^2} \tag{12}$$

or on expressing R by (6) and using $v_0/r_0=\omega$,

$$mv_0^2 = \frac{q^2\nu}{2\epsilon v} \tag{13}$$

Comparing this with (1) yields Planck's constant in the form

$$h = \frac{q^2}{2\epsilon v_0} \tag{14}$$

which is identical with (8), Sec. 2.5, since $K = -q/4\pi\epsilon$.

This provides a check on our calculations, but the important point to note is that the correct result (14) was obtained without breathing a word about energy quantization or atomic spectra. It emerged solely from the Maxwell equations (and energy conservation) by (3), and from the electron oscillations, themselves derived from the Maxwell equations, leading to (1) and (2).

Now let us apply the same procedure to the *rectilinear* motion of the oscillating electron, moving with average velocity v_0 and oscillating with a frequency given by (1).

It is nowhere written that the energy of these oscillations, beyond the energy associated with the *mean* velocity and given by (1), must be the same as in circular orbital motion. There is no continuous transition from circle to straight line: as the radius tends to infinity, velocity and energy go to zero. Moreover, the energy ΔE of the oscillations as established in Sec. 2.8 was based on a velocity that ultimately rested on the ratio of radial to transversal forces (Sec. 2.4); but there are no such forces in rectilinear motion. We therefore set the energy of the oscillations (in excess of $\frac{1}{2}mv_0^2$) equal to

$$\Delta E = \tfrac{1}{2}mv_0^2 a\beta^n \tag{15}$$

where a and n are constants to be determined; for orbital motion we found them in Sec. 2.8 to be $a=1$ and $n=4$.

This time we do not have the privileged value r_0, the radius of the orbit, to work with; however, for rectilinear motion we can use the Faraday field by the method of Sec. 2.7. Since the Faraday field has the direction of the acceleration and equals the electric field less the Coulomb field, it follows from (6) and (7), Sec. 2.7, that for an electron accelerated in the x direction

$$\psi = \frac{2E_0 s}{R}\cos^2\theta \tag{16}$$

where

$$s = \frac{\dot{v}R^2}{2c^2} \tag{17}$$

and E_0 in (16) is the Coulomb field. Hence

$$\int \tfrac{1}{2}\epsilon\psi^2 = \frac{\mu q^2 \dot{v}^2 R^2}{8\pi c^2}\int_R^\infty \frac{dr}{r^2}\int_0^\pi \cos^4\theta\,\sin\theta\,d\theta = \frac{3}{20}\frac{m\dot{v}^2R^2}{c^2} \tag{18}$$

where we have used (6); the value of the integral over θ is 2/5. Substituting this in (5) without the radial term, we have

$$\frac{3}{10}\frac{\dot{v}^2 R^2}{c^2} + a\beta^n v_0^2 \sin^2 \omega t = \text{const} \tag{19}$$

or, since the square of the acceleration must equal $\omega^2 v^2 \cos^2 \omega t$,

$$\omega^2 = \frac{10 a v^n}{3 r^2 c^{n-2}} \tag{20}$$

Taking the square root,

$$2\pi\nu = \sqrt{\frac{10}{3}\frac{6\pi m v^{n/2}}{q^2 \mu c^{n/2-1}}} \tag{21}$$

The frequency is by (1) proportional to the square of the velocity; therefore $n=4$, just as in the circular case. On manipulating (21) into the form (1), and setting $1/a = A$,

$$mv^2 = \frac{1}{3a}\sqrt{\frac{3}{10}} q^2 \mu c \nu \tag{22}$$

so that finally by comparison with (1),

$$h = Aq^2 \mu c = 68.51 q^2 \mu c \tag{23}$$

where the cluster of constants in (21), along with our sins in neglecting delay effects and using (17) for variable acceleration, have been absorbed into the constant A, whose numerical value is then found from the known value of h.

However, these sins are too small to account for a constant as large as 69 (if it were a mere correction factor), so it must be concluded that the unknown constant a in (15) equals approximately 12.5.

It would have been nice to derive the numerical value of Planck's constant exactly, using independently measured constants only. (As a matter of fact, the publication of this book was delayed by close to one year in the hope of achieving that.) Nevertheless, the constant A, derived independently or not, *is* a constant; it therefore follows that Planck's constant is associated only with electric charge — an electron charge at that — but not with any mass.

That means that quantum mechanics works because elementary particles have various masses, but only either one elementary charge or none; only a single Planck's constant therefore occurs in the Schrödinger equation and throughout quantum mechanics. If for example, there were fictitious particles with three elementary charges, their behavior would have to be described by a Planck's constant nine times as large, though independent of their masses.

2.10. The Root Problem

Before we complete Part Two by an investigation why the orbiting electron does not radiate (beyond the simple fact that the energy balance calculated in Sec. 2.8 leaves no energy to be radiated away), there are two more important items to be discussed: first, the physical reason for the charge contraction of Sec. 1.6, which Einstein, in my view, mistook for the contraction of all space; and second, the reason why the electron oscillations derived here (apart from leading to the Schrödinger equation) contradict the Einstein theory.

We were led to the quantization of electron orbits by showing that the electron velocity oscillates as the electron traverses a force field — not the coordinates of an observer — and this now enables us to take a fresh look at the root problem of moving charges.

As we saw in Sec. 1.6, Maxwell's equations and the principle of relativity results in the deformation of the equipotentials round a point charge: they change from concentric spheres round a point charge at rest to ellipsoids (oblate spheroids), contracting in the direction of the velocity. The contracted equipotentials correspond to a charge distribution

$$\rho = \frac{1}{\sqrt{1-\beta^2}}\,\rho_0\left(\frac{x}{\sqrt{1-\beta^2}}, y, z\right) = \rho_0\frac{1-\beta^2}{\sqrt{1-\beta^2\sin^2\theta}} \tag{1}$$

where ϱ_0 is the rest charge distribution, and θ is the angle between the radius vector (origin at the point charge) and the velocity. On this, Lorentz and Einstein agreed. The difference between them, as explained in Part One, was the velocity: was the effect associated with the velocity of the charge with respect to the ether or with respect to an observer?

But the root problem, regardless of the ramifications, was this: is this charge contraction something that is really happening when a charge moves, or is it merely something that is perceived by a moving observer? More generally, do forces act independently of any observer, or are they something that depends on the perception of an individual observer in the manner of a Doppler effect or aberration?

Lorentz took the former view, which I believe to be *fundamentally* correct, even though his understanding that the effect-producing velocity was that with respect to an unentrained ether turned out to be erroneous. It also gave no natural physical explanation why charge distributions should contract as they move through the ether. If they met some kind of resistance akin to friction, one would have thought they would be deformed "aerodynamically" like rain drops, not, as in (1), symmetrically and independently of the sign (sense) of the velocity.

In any case, Lorentz's *interpretation* of the velocity in Maxwell's equations, though it satisfied the relativity principle, was refuted by experiment.

However, that did not invalidate the fundamental classical attitude that natural phenomena generally obey observer-independent laws. As we have seen in Part One, the experimental evidence is perfectly consistent with the understanding that the effect-producing velocity of a charge is that with respect to the traversed field rather than with respect to an observer.

But is there a physical explanation for the contraction of charge? Lorentz did not give one, and to Einstein the contraction of charge was merely due to the perception of an observer for whom space itself had changed. But there is, I believe, a good physical reason for the charge distribution to change when it moves through an electric field, and that is a generalized skin effect.

Skin effect is usually understood to mean the concentration of an alternating current at the outer layers of a cylindrical conductor. More generally, we might define it as the tendency of a moving charge distribution to move with respect to the velocity axis – outward, if the charge is negative.

What happens physically is that the magnetic field $\mathbf{B}$ produced by the current acts on its begetter with a magnetic force $q\mathbf{v}\times\mathbf{B}$. Since $\mathbf{B}$ is related to the Coulomb field by (2), Sec. 1.4, it travels with each electron just as the Coulomb field does. With a steady DC current, therefore, there is no force and no skin effect. But when the charges are accelerated, as they are when the current is sinusoidal, they will attempt to cross their own $\mathbf{B}$ lines (at right angles), for they are now subject to a magnetic force whose direction depends only on the sign of the charge, not on the direction of acceleration. For negative charges (electrons), that direction is *outward from the velocity axis.*

As the current filaments move outward, they take their magnetic field with them, and to find the final current density distribution over the cross section of the conductor, one must set up and solve an equation. That is easily done on eliminating the current density $\mathbf{J}$ by the use of Ohm's Law ($\mathbf{J}=\sigma\mathbf{E}$) from the wave equation for $\mathbf{E}$ in a cylindrical conductor; the procedure leads to the well known skin effect equation, yielding a Bessel function of imaginary argument as the resulting current distribution $\mathbf{J}(r)$.

For a solitary moving electron, the same basic principle applies, for since its velocity fluctuates in self-induced oscillations, it represents an alternating current superimposed on the DC component corresponding to the average velocity. If we imagine the electron as a surface-charged sphere, as we have done throughout Part Two, the same type of magnetic force will act on its surface charge, forcing that charge to flow away from the velocity axis toward the great circle perpendicular to the velocity. This will result in a smaller charge density for $\theta=0$, and a larger density for $\theta=\pi/2$, as is the case in (1).

The quantitative derivation, however, is both difficult and superfluous. It is difficult because there is no Ohm's Law to relate the effective current density to the electric field strength, and a substitute relation between the two is not easy to derive. And it is superfluous because no matter how we derive the result, we have nothing but Maxwell's equations to go on; but Maxwell's equations yield (1) easily and directly (see Sec. 1.6), so that any derivation along the lines suggested above would merely provide an unwieldy duplicate.

What matters here is that the generalized skin effect is a genuine physical phenomenon that really happens, not a perception of an observer determined by the velocity with which he travels past the charge. Relation (1), therefore, is a genuine redistribution of charge as it traverses a field: a contraction in unadulterated, Euclidean space that remains just as unaltered as when heat-shrink insulation contracts in it.

2.11. The Schrödinger Equation

The Schrödinger equation (1926) is formally derived from the de Broglie relation, and therefore formally also follows from (9), Sec. 2.2; however, since we found a physical reason for that relation, we will also be able to give the Schrödinger equation a physical meaning – for in contemporary physics it is simply something that works successfully, but stands on its own as a separate law, without connection to macroscopic physics or derivation from its fundamental laws.

It follows from Sec. 2.3 that the Faraday field satisfies the wave equation

$$\nabla^2 \boldsymbol{\psi} - \frac{1}{c^2}\frac{\partial^2 \boldsymbol{\psi}}{\partial t^2} = 0 \tag{1}$$

(This follows either from the wave equation for the vector potential **A**, or simply from the fact that both the total field **E** and the Coulomb field $\mathbf{E}_c$ satisfy it.)

For the Faraday field of the oscillating electron, the solution is harmonic, so that (1) simplifies to

$$\nabla^2 \psi + k^2 \psi = 0 \tag{2}$$

where the scalar ψ is any component of $\boldsymbol{\psi}$, on the understanding that $\nabla^2 \times \boldsymbol{\psi}$ stands for grad div $\boldsymbol{\psi}$ – curl curl $\boldsymbol{\psi}$ (which is automatically true only for rectangular components). The parameter k is the propagation constant, which in the case of an electromagnetic wave propagating with velocity c equals

$$k = \frac{\omega}{c} = \frac{2\pi}{\Lambda} \tag{3}$$

with Λ the free-space wavelength of the propagating wave.

But (3) is clearly not applicable here: not only because in this natural electron oscillation there is no radiation (as will be shown in Sec. 2.12), but also because we know that the Faraday field surrounds the oscillating electron, which is moving with velocity v. In terms of solutions of the wave equation, ψ is a standing wave with respect to the mean position of the electron, which makes it a wave propagating with velocity v with respect to the field that the electron traverses. (That field and the electron velocity are not, of course, limited to the field of the nucleus and the velocity of an electron in orbit; it applies to an electron traversing any field in any path.)

It will be recalled that the general solution of (1) is given by two traveling waves propagating in opposite directions; depending on their amplitudes, the sum of the two waves may propagate with any velocity from zero (when the amplitudes are equal, resulting in a standing wave) to c (when the amplitude of one of them is zero). In general, when the ratio of the two amplitudes is neither one nor zero, the resulting field can be represented as the sum of a standing and a traveling wave, the latter propagating with a velocity other than c, in spite of the factor $1/c^2$ in (1). Thus the harmonically oscillating ψ field surrounding the electron moving with velocity v is also a solution of (2), but its propagation constant, instead of (3), is

$$k = \frac{\omega}{v} = \frac{2\pi}{\lambda} = \frac{2\pi mv}{h} \tag{4}$$

where we have used the "carrot formula"

$$v = \nu\lambda \tag{5}$$

and the constant of proportionality

$$h = mv\lambda \tag{6}$$

both derived in Sec. 2.2.

Substituting (4) in (2), we have

$$\nabla^2\psi + \frac{4\pi m^2 v^2}{h^2}\psi = 0 \tag{7}$$

But what is usually known about an electron traveling through an external field (not necessarily that of an atomic nucleus) is not its velocity, but its potential and total energy in the entire system — that is, its total energy in the traversed field. We therefore eliminate the velocity by using (24), Sec. 2.1,

$$\tfrac{1}{2}mv^2 = E - U \tag{8}$$

where E is the total energy and U the potential energy of the electron in the field that it is traversing.

Substituting (8) in (7) we obtain

$$\nabla^2\psi + \frac{8\pi^2 m(E-U)}{h^2}\psi = 0 \tag{9}$$

which is the Nobel-prize winning Schrödinger equation.

In quantum mechanics, which is based on the unexplained validity of (9), ψ is the "wave function," whose physical interpretation is that its square, when multiplied by an element of volume, is proportional to the probability of the electron being located in that volume.

But what we have just shown is that the Schrödinger equation is obeyed by the Faraday field — the rotational part of the electric field, or the electric field other than the Coulomb field, as discussed in Sec. 2.3.

I chose the letter ψ for this field because I believe that it is proportional to, and possibly identical with, the wave function in quantum mechanics. The square of the Faraday field is proportional to its energy density, and when multiplied by an element of volume it gives the energy in that volume, which does indeed increase as we approach the location of the electron, as evident from both Secs. 2.3 and 2.10. This formally agrees with the quantum-mechanical interpretation of ψ, yet represents a considerable difference in physical insight. (If someone unacquainted with Western folklore asks about Snow White and is told it is something with a high probability of being in the neighborhood of the Seven Dwarfs, he has not been given false information — but what has he been told about the nature of Snow White?)

The fact that it is the Faraday field that obeys the Schrödinger equation allows a resolution of the two-slit paradox and refutes the "wave-particle dualism," in which two incompatible concepts are proclaimed identical. [A wave is incompatible with a particle for at least three reasons, not counting the disputed capacity to interfere: 1) a wave can be split into two or more parts; 2) it does not repel or attract other waves; and 3) it is attenuated, even in a lossless medium, by natural dispersion, such as that governed by the inverse square law.]

If ψ is a standing wave with respect to the moving electron, then clearly this wave will pass through both slits, though the electron passes only through one. It can be shown, I believe, that the electromagnetic waves re-radiated by the two slits will slow to the group velocity in the neighborhood of the electron, which itself becomes the prisoner of that field. The position of the electron striking the screen is then determined by the interference field of the two electromagnetic waves, which have been slowed to have a de Broglie wavelength rather than a free-space wavelength — just as observed in experiments.

The reason why I have sketched the explanation of the double-slit paradox only qualitatively is that the corresponding mathematical derivations are not as clean as I would like, and this would invite criticism in an area far removed from the Einstein theory. Thus the objections to my interpretation of double-slit electron diffraction, whether merited or not, would distract from the main object of this book: the replacement of the Einstein theory by a simpler and more rational one.

Returning to the Einstein theory, the question arises whether that theory could not also derive the results of Part Two — the de Broglie relation, the Bohr orbits, the Schrödinger equation, the insight into Planck's constant, and the various verifications of the present theory's consistency. The aberration of force and the resulting modified Newton Law, for example, which contradict the Einstein theory, were not used in the derivation of the electron path in Sec. 2.4; and as shown in Sec. 2.9, an alternative, purely electromagnetic derivation is also possible. Could the Einstein theory, then, not do equally well?

I do not believe so. What all of Part Two rests on are the electron oscillations explained in Sec. 2.2 and derived in Sec. 2.3. They result from the Maxwell equations, or more particularly, from Ampère's, Faraday's and Lenz's Laws – under the implicit assumption that the effect-producing velocity of the moving electron is that with respect to the traversed field.

But if, as postulated by the Einstein theory, the velocities that make the Maxwell equations valid are those with respect to an observer, does the electron stop oscillating when the observer is at rest with respect to it? Does its frequency and amplitude of oscillation depend on the velocity with which the observer moves past it? If so, it would not be a Doppler-like change in frequency that shifts a basic frequency by a ratio resulting from the Lorentz transformation; it would be *some* frequency for the moving observer and none for the stationary one.

But this is not just inconsistent: it kills the relativity principle. The Einstein theory rescues the relativity principle by adjusting space and time: the two observers in different inertial frames observe the same laws of physics because the change in these laws is exactly compensated by the change in space and time coordinates. Unless the velocities of the two frames differ by that of light, these changes and compensations amount to only a part of the corresponding quantity (as dictated by the Lorentz transformation), not to its total annihilation. But in the case of electron oscillations, the Einstein theory would have the observer traversing an electric field with an electron in his pocket see no oscillations at all; yet that same electron would oscillate for an observer at rest with respect to that same traversed field. The two observers in the two inertial frames thus register different laws of physics (beyond repair by the Lorentz transformation), and the relativity principle cannot hold.

It would thus appear that the Einstein theory, which sacrificed the conventional concepts of space and time because of its faith in both the Maxwell equations (with observer-referred velocities) and the relativity principle, leads to a contradiction between the latter two.

In the present theory, conventional (observer-independent) space and time, the Maxwell equations and the relativity principle can all live together without contradiction; the flaw is in the Einstein theory's assumption that the laws of physics apply with observer-referred rather than field-referred velocities.

2.12. Radiation and Some Other Matters

We have seen that the self-induced oscillations of the electron in the Faraday field lead to the Bohr orbits as the only stable ones in quasi-central motion, i.e., in motion slightly deviating from central due to the aberration of force.

Maxwell's equations with field-referred velocities thus yield the same *observable* results as Bohr's postulates. But in interpretation, the present theory differs significantly from that of modern physics. The electron is in the stable orbits predicted by Bohr and quantum mechanics, not because of any postulates assuming behavior unknown in the macroworld, but because they are the only orbits where the energy of the electron is in balance: not just the mechanical energy (kinetic and potential), but the sum of mechanical and electromagnetic, with the oscillation frequency a harmonic of the angular frequency. If this balance is upset by an external force, the electron will spiral along a predictable path to a lower or higher stable orbit where that double balance is once more restored.

As the electron spirals to another orbit, it follows the same laws as a satellite firing a rocket that will take it to a higher or lower orbit; the satellite, it is my conviction, would also settle only in certain discrete stable orbits if it could be given a correspondingly large electric charge. [The scaling would have to be done by (6), Sec. 2.6, showing the charge required for the satellite to be far too large to make such an experiment realistic.] The laws of physics need no longer be balkanized into the laws of the "micro" and "macro" worlds, with only the tenuous link of the Correspondence Principle to connect them. There is only one physics in the proposed theory, and I consider this aspect its most valuable asset.

Now to another point: how quantized is energy?

First of all, it never *was* quantized like electric charge or uncracked eggs. These come as integral multiples of one standard unit, but the energy quantum $h\nu$ assumes a value from a continuous range with no upper bound. However, even this property now appears less striking, for it is based on no more than the harmonics of a natural frequency, such as that of a violin string or a window pane—or to make the analogy more fitting to a nonlinear system, such as that of a tuned circuit with an iron-core inductance. Harmonics are, of course, by definition integral multiples of a fundamental frequency, and subharmonics integral parts of it.

Viewed as a resonator, a charge oscillating in its own Faraday field is a *nonlinear* system, for the differential equation governing its behavior, such as (15), Sec. 2.4, is nonlinear [we approximated it by a linear equation by regarding (14) of that section as a constant.]

Linear systems generally oscillate at a fundamental (natural) frequency ω as well as its higher harmonics (2ω, 3ω, ...); nonlinear systems, in addition, produce subharmonics with frequencies $\omega/2$, $\omega/3$,... Not all harmonics or subharmonics, however, are necessarily produced in the oscillations: attenuation or total suppression of certain harmonics is what gives musical instruments their characteristic timbre. A *flageoletti* tone on a violin completely suppresses the odd harmonics (because the violinist very lightly touches, rather than firmly clamps, the string at one half or quarter its length).

Quite similarly, the self-induced oscillations of an orbiting electron have a fundamental frequency (12), Sec. 2.9, plus subharmonics as given in Sec. 2.5, with frequencies that are an integral part of the fundamental: the orbital frequencies divide the fundamental by n^3, and the "wiggle" frequencies divide it by n^2. (The wiggle frequency $n\omega$ is an *harmonic* of the orbital frequency ω_n, which itself is a *subharmonic* ω_1/n^3 of the ground-level orbital frequency ω_1.) Subharmonics with other powers of n, as well as all higher harmonics, are "suppressed," meaning that they do not satisfy the stability criteria.

Finally, there is the question of why the electron in stable orbit, though it moves under constant acceleration, does not radiate.

The reason is that **E** and **H** are in phase quadrature, so that the Poynting vector reverses direction twice per cycle as the energy flows from the magnetic-kinetic energy ($\frac{1}{2}mv^2$) of the electron into the Faraday field and back, precluding any radiation in a constant direction.

This will be shown in more detail below, but first let me reject the widely held belief that an accelerated charge *must* radiate. This belief is unfounded, for accelerated charges that do not radiate are commonplace. True, there is no known case of electromagnetic radiation that is not due to accelerated electrons. But the converse is certainly not true — there are electrons that are accelerated millions of times a second without radiating a picowatt: in a transmission line such as used to feed a transmitter antenna in the microwave band, for example. They are accelerated just as much as those in the antenna, but if the characteristic impedance of the line is matched to the radiation resistance of the antenna, they will not radiate any more than their cousins in the atom.

There is a possible, though remote, similarity between the absence of radiation in a transmission line and its absence for the orbiting electron. The electrons oscillating in a transmission line do produce an electromagnetic wave, but it propagates along the line without radiating until it strikes the (matched) antenna at the end of the line, and only then does it take off. If that discontinuity is eliminated (by making the line infinite or terminating it with a matched heat-absorber), there is no radiation at all.

But a transmission line is mathematically equivalent to a chain of an integral number of identical four-poles with lumped circuit elements. It will not radiate if it is infinite or terminated by a matched impedance. If such a chain is turned back onto itself and made lossless (by cooling it below the superconductivity level, for example), then after initially energizing it, it should have an electromagnetic wave going round and round, interacting with the accelerated electrons, and radiating exactly nothing — by orthodox, old-fashioned transmission line theory without any additional assumptions of any kind. (To my knowledge, such an experiment has never been performed. The main difficulty, starting the wave off in only one direction, can probably be overcome, and thereafter the wave should keep going for hours, perhaps even days or weeks.)

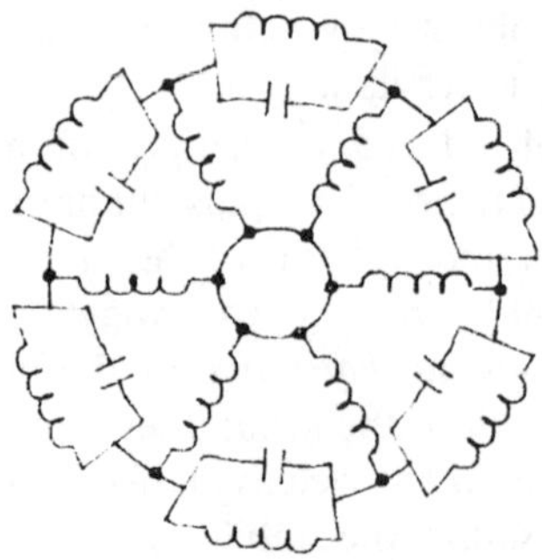

Circular chain of identical, lossless fourpoles will not radiate.

The similarity with electron orbits is this: there will be no radiation when the ring of lumped four-poles has an integral number of identical elements. But if the number is non-integral (i.e. one of the four-poles contains only a fraction of the "regular" parameters in one lumped circuit element), the resulting discontinuity will produce a reflection and hence a standing wave, and the ring *will* radiate. Similarly, an electron in an orbit with a non-integral number of wiggles will have the energy balance disturbed, and the excess energy will be radiated (or the deficit absorbed) until the electron reaches a stable orbit with an integral number of undulations.

One might also think of the two cases in terms of the presence or absence of discontinuities: when the transmission line wave strikes the antenna, it has nowhere to go but into space, and when some outer force knocks the electron out of its orbit, it has only radiation to dispose of its excess energy.

The refusal of the orbiting electron to radiate marred Lord Rutherford's early model of the atom, and the postulates of his pupil Niels Bohr threw no light on the subject; in fact, Bohr had to introduce a special postulate to keep the electron in stable orbit from radiating. But not all is well with such a postulate, for the Correspondence Principle (requiring the transition to classical physics when the number of particles becomes very large) will not work on it. If one accelerated electron does not radiate, how about two, or three – or 278 as in the uranium atom? Or a trillion uranium atoms?

But the present theory does not need a Correspondence Principle, for the same laws apply to all of physics. Electrons in atomic orbit, or other charges moving with non-uniform velocity, can be accelerated without radiating if they give rise to a Poynting vector that periodically changes direction – in other words, when the energy flow oscillates, but does not proceed forward in the same direction (of propagation), and when the energy balance is achieved by non-radiative energy forms. The electron in orbit does not radiate because its **E** and **H** fields are in phase quadrature, so that the Poynting vector does not have the constant direction of propagating radiation, but changes direction twice per cycle; the energy flows from the field surrounding the electron to the electron itself as it accelerates, and back again as the Faraday field decelerates it.

This latter case arises in the *natural* oscillations of an electron, where the energy balance of Faraday and magnetic field automatically introduces a phase quadrature of the vectors entering the Poynting vector product; but it does not arise in the *forced* oscillations (such as in a radio antenna) using external energy, which is then radiated away and associated with a Poynting vector with an active, in-phase component.

Bohr had no Poynting vector to consider, because he postulated the physics of the orbits away. He postulated the *result* correctly, but forewent the insight offered by its derivation – an early case of derailing physics from understanding to description.

Now to the matter of radiation. Let us first note that the Poynting-Heaviside Theorem,

$$\iint_S (\mathbf{E} \times \mathbf{H}) \cdot d\mathbf{S} = -\iiint \mathbf{E} \cdot \mathbf{J}\, dV - \frac{\partial}{\partial t} \iiint (\tfrac{1}{2}\epsilon E^2 + \tfrac{1}{2}\mu H^2) dV \tag{2}$$

can be simplified if, as before, we set the electric field $\mathbf{E}$ equal to the sum of the Coulomb field $\mathbf{E}_c$ and the Faraday field $\boldsymbol{\psi}$, for in our case the Coulomb field makes no contribution to (2): the directions of $\mathbf{E}_c$, $\mathbf{H}$ and $d\mathbf{S}$ are, respectively, $\mathbf{r}_0$, $\mathbf{r}_0 \times \mathbf{v}_0$ and $\mathbf{r}_0$, where the subscript 0 denotes unit vectors, so that the contribution of the Coulomb field to the integrand on the left contains the factor

$$[\mathbf{r}_0 \times (\mathbf{v}_0 \times \mathbf{r}_0)] \cdot \mathbf{r}_0 = [\mathbf{v}_0 - (\mathbf{r}_0 \cdot \mathbf{v}_0)\,\mathbf{r}_0] \cdot \mathbf{r}_0 = 0 \tag{3}$$

The direction of the effective current density $\mathbf{J}$ is $\mathbf{v}_0$, so that the dot product in the integrand containing it is proportional to $\cos\theta$, which makes the volume integral vanish when integrated over θ from 0 to π, and finally in the last term, the energy associated with the Coulomb field vanishes because it is constant in time. Therefore (2) simplifies to

$$\iint_S (\boldsymbol{\psi} \times \mathbf{H}) \cdot d\mathbf{S} = -\iiint \boldsymbol{\psi} \cdot \mathbf{J}\, dV - \frac{\partial}{\partial t} \iiint (\tfrac{1}{2}\epsilon \boldsymbol{\psi}^2 + \tfrac{1}{2}\mu H^2) dV \tag{4}$$

The Poynting vector is then by (4) and (12), Sec. 2.6,

$$\boldsymbol{\psi} \times \mathbf{H} = \frac{\mu q^2 \sin^2 \theta}{(4\pi)^2 r^3} \dot{v} v \mathbf{p}_0 + \cdots \tag{5}$$

where

$$\mathbf{p}_o = [\dot{\mathbf{v}} \times (\mathbf{v}_0 \times \mathbf{r}_0)]_0 + \cdots \tag{6}$$

In the last two expressions, the three dots show that we have ignored the velocity component in (12), Sec. 2.6, in anticipation of the integration, which makes the corresponding terms vanish.

The important point here is the product $\dot{v}v$in (5), for $\dot{v}$ and v are in phase quadrature so that the Poynting vector will change direction twice per period as the energy flows from the kinetic energy of the electron to the surrounding field and back. This is quite unlike radiation where ψ and **H** are in phase and therefore reverse directions simultaneously, leaving the Poynting vector aligned in the direction of propagation.

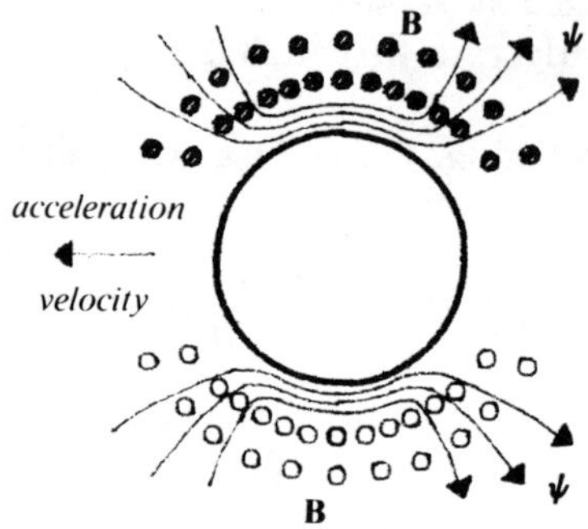

Field of an accelerated electron. Full dots show **B** going into the paper; arrows show the direction of the Faraday reaction opposing the acceleration (regardless of the conventional direction of lines of force, which depends on the sign of the charge).

It may be helpful to draw a field sketch to gain insight into the energy flow. If the velocity is directed as shown in the figure, the magnetic field lines will ring the electron sphere, into the paper at the top, and most densely at the top and bottom due to the $\sin\theta$ factor. The lines of the Faraday field ψ curl round this magnetic field (their density proportional to the rate of change of **B** in time) and close on themselves, for by (19), Sec. 2.3, the Faraday field has no divergence.

The Poynting vector at the surface of the sphere is shown in the figure on the next page. Since both **H**, which curls round the velocity vector, and ψ, which curls round the $\times\,\partial\mathbf{H}/\partial t$ vector, are tangential to the sphere, the Poynting vector must by (6) point radially outward or inward: outward when the electron is decelerating and the energy is flowing into the field, and inward when the Faraday force accelerates the electron, increasing its kinetic energy at the expense of that of the field.

Now consider the energy balance (4). The only term that has not yet been calculated in previous sections is the term

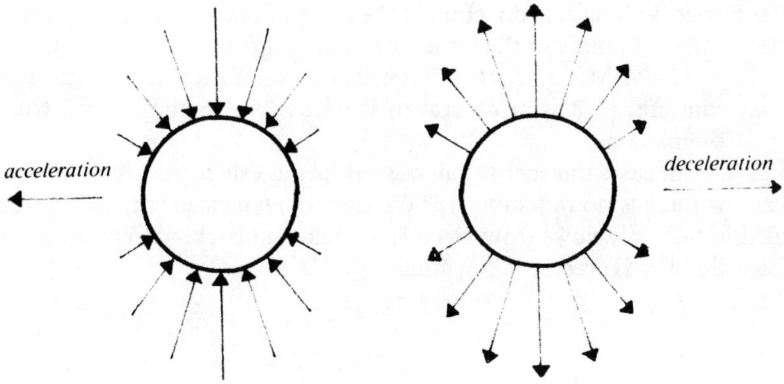

Energy flow (Poynting vector) to and from fiield.

$$\int\iiint \boldsymbol{\psi}\cdot\mathbf{J}\,dV\,dt = \int\iiint \rho\mathbf{v}\cdot\boldsymbol{\psi}\,dV\,dt = \int q\mathbf{v}\cdot\boldsymbol{\psi}\,dt = -\int \frac{qv\dot{v}\phi}{c^2}\,dt$$
$$= -\int \frac{qv\phi}{c^2}\,dv = -\tfrac{1}{2}q\beta^2\phi = -\tfrac{1}{2}mv_0^2 \tag{7}$$

where the last equality is based on the relation $q\phi = mc^2$, explained after (21), Sec. 2.3.

Thus the energy balance (4) amounts to

$$-\tfrac{1}{2}mv_0^2 = \tfrac{1}{2}mv_0^2 - \tfrac{1}{2}mv_0^2 - \tfrac{1}{2}mv_0^2 \tag{8}$$

where the order of the terms is the same as in (4). The total energy of the moving electron is therefore

$$mv_0^2 = mv_0\lambda\nu = h\nu \tag{9}$$

in agreement with our original derivation in Sec. 2.2.

This case of natural oscillations, when nothing is radiated because the energy flows from the oscillating electron to its field and back, should be contrasted with the case of *forced* oscillations, when the electrons oscillate in an impressed **E** field, as they do, for example, in a transmitter antenna. If we leave out the terms corresponding to (4), which add up to zero, and assume a lossless antenna ($\mathbf{E}\cdot\mathbf{J} = 0$), the Heaviside-Poynting Theorem becomes simply

$$\iint_S (\mathbf{E}\times\mathbf{H})\cdot d\mathbf{S} = -\frac{\partial}{\partial t}\iiint (\tfrac{1}{2}\epsilon E^2 + \tfrac{1}{2}\mu H^2)dV \tag{10}$$

where the externally impressed $\mathbf{E}$ imparts a velocity to the free electrons in the antenna, thus endowing them with a magnetic field strength $\mathbf{H}$ that is *in phase* with it. Hence both reverse direction at the same moment and the Poynting vector on the left always points in the same direction — the direction of radiation away from the antenna. Moreover, the frequency of oscillation is not determined by Planck's constant, as it is in natural oscillations, but by the Federal Communications Commission.

Thus in both cases the energy balance works out exactly as it should; in natural oscillations there is no radiation, for the electromagnetic energy that "should" be radiated in fact only flows from electron to field and back, always in the direction in which the $\mathbf{E} \times \mathbf{H}$ vector is poynting.

Part Three

Einstein Plus Two

3.1. Gravitation

If forces are propagated from their source with a finite velocity c and therefore act on bodies at a distance with the corresponding delay and aberration, then the modified Newton-Coulomb Law of Sec. 1.8 must hold for Newton's as well as for Coulomb's Law. If the force is close to perpendicular to the velocity,

$$\mathbf{F} = \frac{\Gamma M_1 M_2}{r^2(1-\dot{r}/c)^2}[(1-\beta^2)\mathbf{r}_0 + \beta\mathbf{\Theta}] \tag{1}$$

where $\Gamma = -6.67 \times 10^{-11}$ Nm²/kg² is the gravitational constant, and where, as explained in Sec. 1.8, the additional term in the $\mathbf{\Theta}$ direction is the aberrational component of the force, the parenthesis in the denominator stems from the delay, and the expressions in $\beta = v/c$ are second-order approximations for $\cos\beta$ and $\sin\beta$, respectively; as always with polar coordinates, β and v refer to the transversal (not radial) component of the velocity only.

The gravitational force between two equal masses M at rest with respect to each other is thus

$$\mathbf{F} = \frac{\Gamma M_1 M_2}{r^2}\mathbf{r}_0 \tag{2}$$

which equals the instant-action-at-a-distance (IAAD) expression, as it must.

The idea of gravity propagating from its source with a finite velocity is often thought to be due to Einstein.

Not so: The idea stems from none other than the incomparable Pierre Simon Marquis de Laplace, who considered it in his *Traitise sur la théorie de la mécanique céleste*, begun in 1799, the last volume published in 1825. Chapter VII, Book 10, is entitled "On the alterations which the planets and comets may suffer by the resistance of mediums they pass through, or by the successive transmission of gravity," and section 22 (pp.642-645 of the English translation [1839/1966]) considers the case of "gravitation produced by the impulse of a fluid directed toward the center of the attracting body." Laplace used the word "aberration" and calculated its value; however, from the data then available he reached the conclusion that gravitation must propagate at least 100 million times faster than light, noting that therefore mathematicians are justified in considering its velocity infinite – an assumption that Newton made without such calculations, but not without misgivings.[1]

Now let us recall a 75-year old discovery that has been all but forgotten. In 1912, Yale physics professor Leigh Page derived the Maxwell equations by applying the Lorentz transformation to the Coulomb Law without any further assumptions. To me that means that the Lorentz transformation is a way of arriving at correct results by compensating, at the price of distorting space and time, for an inaccurate Newton-Coulomb Law. But more important at the moment is another point: *Since the Lorentz transformation is not clever enough to know what the symbols stand for, it follows that the forces and fields of two masses must obey the Maxwell equations in the same way as two opposite charges in electromagnetics.*

To stress analogous quantities and procedures, we will continue to measure inertial mass in kilograms, but gravitational mass will be expressed in "coulomb equivalents" defined as the electric charge that two gravitational masses would have to carry in order to experience the same absolute value of force attracting them. (The absolute value eliminates repulsion, i.e. bridges over the dissimilarity between possibly negative charge and always positive mass.) Applying this definition to the force between two equal gravitational masses M at rest, we have

$$|\mathbf{F}| = \frac{\Gamma M^2}{r^2} = \frac{Q^2}{4\pi\epsilon_0 r^2} \tag{3}$$

where Q is the equivalent electric charge, or the gravitational mass expressed in coulomb-equivalents. Hence

$$Q = M\sqrt{4\pi\epsilon_0\Gamma} = 8.613 \times 10^{-11} M \tag{4}$$

i.e., each kilogram of mass is the equivalent of 8.613×10^{-11} coulombs producing the same attraction. To stress the analogy between electromagnetic and gravitational fields, let us precede the gravitational quantities by the superscript 0 to remind us that the quantity is associated with *uncharged* matter, measured in electrical units by assuming an equivalent charge that would produce the same force. In most cases – in *all* cases considered in this book – the gravitational field becomes negligible when a *genuine* electromagnetic field is present.

Under this convention, the "gravi-electric" field is

$${}^0\mathbf{E} = \frac{\mathbf{F}}{Q} \tag{5}$$

[1] "[T]hat one body," he wrote, " may act upon another at a distance through a vacuum, without the mediation of any thing else, by and through which their action and force may be conveyed from one to the other, is to me so great an absurdity, that I believe no man who has in philosophical matters a competent faculty of thinking, can ever fall into it." (Quoted without reference in B. Hoffman, *Relativity and its Roots*, Freeman, New York, 1983.)

where ${}^0Q$ is the "gravicharge" of a small test-mass just as in electrostatics, and the "gravimagnetic" field of a moving particle (or body reduced to an equivalent particle by the Divergence Theorem) is then described by the "gravimagnetic" flux density

$$ {}^0\mathbf{B} = \frac{\mathbf{v} \times {}^0\mathbf{E}}{c^2} \tag{6} $$

With this change in units and the definitions (5) and (6), the Maxwell equations (3) to (6), Sec. 1.10.1, must hold for the reason given above — at least for two particles (or corresponding bodies) where algebraic signs implying the repulsion of charges or currents can simply be changed.

The constituent equations will hold, too, but are not needed: gravitational force is not modified inside matter in the way electric force is altered (there are no known materials wholly or partially opaque to gravity), so that ϵ and μ do not deviate from their free-space values, and the vectors ${}^0\mathbf{D}$ and ${}^0\mathbf{H}$ are physically superfluous. The only essential difference from genuine electromagnetics arises in the algebraic sign of force: care must be taken to get the equivalent sign of the charge right, as in the Lorentz force (1), Sec. 1.10.1 (which we shall not need).

Now let us look at a fundamental point that contradicts accepted physics: the equivalence of inertial and gravitational mass. Many readers will doubtlessly be reluctant to tamper with this ingrained equivalence, but it should be remembered that the only question of importance in a contradiction is not whether it contradicts tradition and authority, but whether it contradicts the experimental evidence.

To compare inertial and gravitational mass (that is, the quantity of matter as measured by its resistance to change of momentum and that quantity as measured by its attraction by other matter), we note that the left side of (1) is the rate of change of momentum, involving inertial mass which, as follows from the Maxwell equations (see Sec. 1.7), is velocity-dependent. On the other hand, gravitational mass, being the equivalent of electric charge, is invariant and unchanged by the velocity dependence of aberration, which only shifts the angle of arrival of the force, but does not change its absolute value. This is evident by writing (1), corresponding to (8) and (9), Sec. 1.8, as

$$ |F| = \frac{K}{r^2}\sqrt{1-\beta^2} \tag{7} $$

or more explicitly,

$$ \frac{m_0}{\sqrt{1-\beta^2}} \left.\frac{dv}{dt}\right|_{\perp} = \frac{\Gamma M_1 M_2 \sqrt{1-\beta^2}}{r^2} \tag{8} $$

On the left of (8) we have the rate of change of momentum, involving the inertial mass with its velocity dependence; on the right we have a velocity dependence that, as we know from Sec. 1.8, is due in part to the bunching of the field, in part to the equivalent of the magnetic field produced by a charge — here a mass — in motion. It would be possible to incorporate the square root in a variable gravitational mass by the relation

$$M = M_0\sqrt{1-\beta^2} \quad [?] \tag{9}$$

but such a procedure would lead to unredeemed confusion, for the velocity dependence of a gravitational mass thus defined would be different from the velocity dependence of inertial mass. The far more natural and convenient way is to regard gravitational mass as invariant in analogy to electric charge, and attribute the velocity dependence factor to its physical causes — bunching of the field and (gravi)magnetic component — exactly as we did with electric charge in modifying Coulomb's Law.

But no matter how we interpret (8) — amounting to no more than the names we give to its factors — it is clear that inertial and gravitational mass can at best be equivalent at rest, when (2) is valid.[1]

The experiments showing the equality of inertial and gravitational mass have always been performed at $\beta = v/c$ near zero (v is the velocity of one mass in the field of the other), utterly failing to provide a test of (7). Typically, the well known experiments by Eötvös and Zeeman, comparing the weight of bodies with their inertial reaction to the centrifugal force on the rotating earth, did not look for a dependence on velocity and could not have found one if they had, for in moderate latitudes the rotational velocity of the earth's surface results in $\beta^2 \approx 10^{-12}$. Not only was this beyond the accuracy of the experiments, but there was no provision to compare the results at different velocities (latitudes). Nor does the advance of Mercury's perihelion provide a check, for as we shall see in the next section, this rests almost entirely on the delay effects due to the *radial* velocity of the planet. In effect, the equality of inertial and gravitational mass has been experimentally demonstrated only when the two attracting masses are at rest with respect to each other, which agrees with (2), (7) and (8).

Neither does the gravimagnetic field $^0\mathbf{B}$ in the "gravitational" Maxwell equations contradict experimental observations, for it is so weak that it is difficult to imagine a laboratory experiment that could directly demonstrate it. As pointed out in Sec. 1.7.2, even the magnetic field in *electro*magnetics is so weak that it becomes readily observable only after the electric field has been neutralized — when charges flow in a conductor that is electrically neutral as a whole. Yet an electric field is vastly stronger than a gravitational one when the comparison is made for the same mass in a charged or uncharged state. It follows from (4) and (6) that the gravitational field strength $^0\mathbf{E}$ of a kilogram of mass is itself very weak, and a glance at (6) shows that this weak field begets an ultraweak

[1] The "at best" refers to recent suggestions that gravitational mass may depend on the material constitution of the attracted body, see [Fischbach and Aronson 1986].

gravimagnetic field $^0\mathbf{B}$ that cannot be expected to have any effect except when masses of astronomic size move with velocities commensurate with the velocity of light. That is the case for the planets, whose β is of the order 10^{-4}.

For orbital motion, the same principles apply as those discussed in Part Two for the case of an electron orbiting a proton: while for the original IAAD Newton-Coulomb Law *any* ellipse represents a possible orbit because the motion is strictly central, the force given by (1) is not strictly central due to the transversal component. This will lead to unstable orbits by higher-order, uncanceled terms of aberration and the Biot-Savart force (see Sec. 1.8) unless the transversal component is somehow neutralized.

However, the type of oscillations leading to the Bohr orbits cannot work for the planets, whose large kinetic energy cannot possibly be counterbalanced by the tiny Faraday field. The charge-to-mass ratio of uncharged matter, given by (4), is a mere 4.9×10^{-22} times that of an electron, making uncharged matter far too clumsy to be stabilized in Bohr orbits: the natural frequencies of the planets found from the equivalent of (5), Sec. 2.9, is several orders removed from their orbital frequencies determined by their distances from the sun.

This appears to leave only one other possibility of orbits that are not destabilized by the transversal component of the field, namely those in which this component is simply absent. The transversal (θ) component satisfies the wave equation (derived from the Maxwell equations), and it is therefore absent at the zeroes of the solution of that wave equation.

To repeat this with a little more detail: since the "gravielectric" field must satisfy the Maxwell equations, it must also satisfy the wave equation derived from them,

$$\nabla^2(^0\mathbf{E}) - \frac{1}{c^2}\frac{\partial^2(^0\mathbf{E})}{\partial t^2} = \mu\, ^0\mathbf{J} \tag{10}$$

where the right side will equal zero just outside the moving planet. (The solution of this equation need not, of course, be a propagating "radiation" wave.) Multiplying by ^0Q yields the wave equation for the force vector attracting the planet; for the θ component of that force in (1), we therefore have

$$\nabla^2\mathbf{F}_\theta(r,\theta) - \frac{1}{c^2}\frac{\partial^2\mathbf{F}_\theta(r,\theta)}{\partial t^2} = 0 \tag{11}$$

The zeroes of the solution of this equation in polar coordinates indicate the only distances r from the center (such as the sun) at which the transversal component vanishes, and for which the orbit is therefore stable.

In Sec. 3.3 we will find these zeroes: they form a discrete set, namely a geometric series in the distances from the center, which is in agreement with the observations on the four planetary systems known to us.

But first a look at Mercury.

3.2. Mercury

It was observed as early as 1880 that the axis of Mercury's elliptical orbit is turning very slowly in the direction of its rotation about the sun. The phenomenon is known as the advance of Mercury's perihelion (the latter being the point on the orbit closest to the sun). The rate at which the axis is turning is a tribute to the astronomers for being able to detect, let alone measure, it: about 43 seconds of arc (0.012 of one degree) *per century*. For Venus, Earth, and Mars, the rate of rotation, again in seconds of arc per century, is 8.6, 3.8, and 1.35, respectively. Actually the astronomers are even better than that, for what they can directly observe are these angles multiplied by the orbit's eccentricity, and even for Mercury, the only one of these to have a significant eccentricity, this makes a minuscule 8.82 seconds of arc per century. (The figure on the right exaggerates the advance $\Delta\theta$ by a factor of about 540,000).

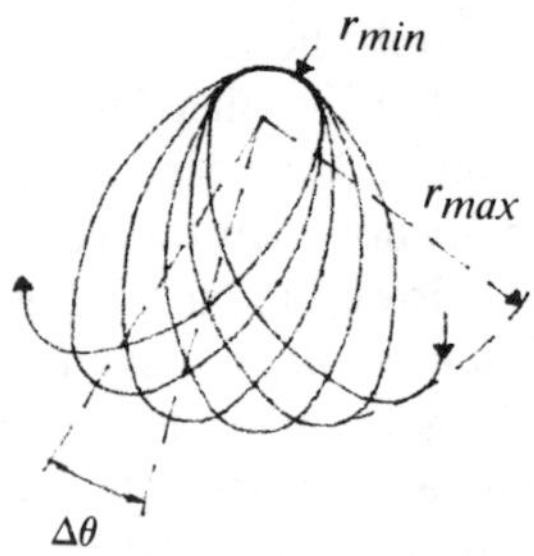

Advance of Mercury's perihelion; $\Delta\theta$ magnified 540,000 times.

Mercury's motion contradicts Newton's original law of gravitation, for if the field were strictly central, and strictly inverse-square, the orbits (if they are finite) would have to be closed curves. It is shown in theoretical mechanics that there are only two types of central field which give rise to closed orbits: those whose potential is proportional to $1/r$ (such as the traditional Newton-Coulomb field) and those where it is proportional to $1/r^2$ (but not, for example, to the sum of these two). All other central fields result in orbits whose axis rotates by an angle $\Delta\theta$ per orbit, as shown in the figure.

Based on his general theory of gravitation and acceleration published in 1915, Einstein [1915] derived the formula

$$\Delta\theta = \frac{6\pi\Gamma M}{ac^2(1-\epsilon^2)} = \frac{24\pi^3 a^2}{c^2T^2(1-\epsilon^2)} \tag{1}$$

where ϵ is the eccentricity of the orbit, a is the semimajor axis, and T the period [the second expression follows from (18), Sec. 2.1, last expression]. This formula is in very good agreement with observations of Mercury, which is the only planet allowing reliable comparison: for reasons explained in numerical astronomy, the observable quantity is not $\Delta\theta$, but $\epsilon\Delta\theta$, and only Mercury's eccentricity $\epsilon = 0.2056$ is big enough to make this product 8.82 seconds of arc per century; the eccentricities of the orbits of Venus, the Earth, and Mars are 0.0068, 0.0167, 0.0034, so that the observable values would only be 0.05, 0.07, 0.13 seconds of arc per century, respectively.[1]

[1] The latter values are taken from G.A. Chedotarev, *Analysis and Numerical Methods of Celestial Mechanics*, Amer. Elsevier Publ. Co., New York, 1967, p. 71. All other astronomical constants in this section are taken from *ABC der Astronomie*, Brockhaus, Leipzig, 1960.

But once again, Einstein was not the first to derive the Mercury formula (1). It had been derived 17 years earlier by Paul Gerber [1898] by classical physics using the same assumption that I am using now – the propagation of gravity with velocity c. For readers who find that hard to believe, Gerber's final expression is reproduced here:[1]

kann. Man erhält daher schliesslich

$$c^2 = \frac{6\pi\mu}{a(1-\varepsilon^2)\psi}.$$

Hierin ist

$$\mu = \frac{4\pi^2 a^3}{\tau^2},$$

wenn τ die Umlaufszeit des Planeten bedeutet. Speziell für Merkur gelten folgende Werte:

$$a = 0{,}3871 \cdot 149 \cdot 10^6 \text{ km},$$

$$\varepsilon = 0{,}2056,$$

$$\tau = 88 \text{ Tage},$$

$$\psi = 4{,}789 \cdot 10^{-7}.$$

Man findet damit

$$c = 305500 \text{ km/sec.}$$

Gerber's derivation of what is now known as "the Einstein formula" in 1898 (*Zeitschrift für Mathematik und Physik*, vol. 43, pp.93-104; the sample above is from p. 103). Gerber inverted the final formula to determine the velocity of light c; his ψ is our $\Delta\theta$, the advance of Mercury's perihelion.

Gerber started from the delayed potential to terms down to $1/c^2$,

$$U = -\frac{K}{r(1-\dot{r}/c)^2} = -\frac{K}{r}\left(1 + \frac{2\dot{r}}{c} + \frac{3\dot{r}^2}{c^2}\right) \tag{2}$$

and from there took a long and arduous road to derive (1). He was presumably not familiar with what a small perturbation in a potential will do to an orbit in

[1] Gerber appears to have been a high-school science teacher in the little German town of Stargard, Pomerania (now Poland, near Szeczyn). His 1898 paper was an extract from a longer work, published in 1902 as a report by his high school (*Realgymnasium*). Since the report was not easily accessible, yet considered important, the full 1902 report was reprinted posthumously in *Annalen der Physik* in 1917 [Gerber 1902/1917].

The town of Stargard was virtually obliterated in World War II, so it may be difficult to find out more about the man who discovered the mechanism turning Mercury's orbit, and who somehow vanished from history. My *guess* that he was a high-school teacher at the Stargard *Realgymnasium* is only based on the fact that the school published his reports, and that unlike most other contributors to the 1898 volume of *Zeitschrift für Mathematik und Physik*, who are identified by their advanced degrees or academic standing, the author's name is simply given as "Paul Gerber."

general, and neither, for that matter, was Einstein 17 years later: when Einstein [1915] finally emerged from Riemannian geometry and gravitational tensors with an additional term supplementing the Newtonian potential, he solved the resulting equation by an approximation involving elliptic integrals.

From this I take it that in 1915, let alone 1898, the simple and general formula for the advance $\Delta\theta$ (see figure on p. 121) in central motion with a perturbed Newtonian potential was not known. It will considerably shorten both Gerber's and Einstein's derivation, so let me first briefly sketch its derivation, leaning heavily on Landau and Lifshits [1965; see Chapter III, Problem 3 at the end of Sec. 15].

Since a $1/r$ (IAAD) potential leads to a closed ellipse (we assume a negative total energy, excluding conics corresponding to an escape from the planetary system), an advance of the type shown in the figure must be caused by a perturbation of this basic potential. We therefore set the potential equal to a "regular" term plus a small perturbation proportional to other powers of r:

$$U = -\frac{K}{r} + \Delta U = -\frac{K}{r} + \frac{C_n}{r^n} \tag{3}$$

where C_n is a constant, and the second term is small compared with the first.

Then we turn to the basic formula for the orbit (13), Sec. 2.1, substituting this general perturbed potential energy (3) for the special case K/r used there. To find the advance $\Delta\theta$ shown in the figure, we integrate from r_{max} to r_{min} and back again to r_{max}, which is the same as twice the latter trip; and in order to evade diverging integrals in the following, we re-write the integral (13), Sec. 2.1, in the form of a derivative with respect to L. All of this yields

$$\Delta\theta = \frac{\partial}{\partial L}\int_{r_{min}}^{r_{max}} \sqrt{2m(E-U) - \frac{L^2}{r^2}}\; dr \tag{4}$$

Now expand (4) in a power series with respect to ΔU; the constant (zeroeth) term of the expansion is 2π, and the first-order term is the required value $\Delta\theta$, the advance by which the polar angle exceeds a full orbit of 2π. After the expansion has been performed, we change the variable of integration from r to θ, using the unperturbed orbit [see (15), Sec. 2.1]

$$r = \frac{p}{1 - \epsilon\cos\theta} \tag{5}$$

(in this section, ϵ stands for eccentricity), so that we finally obtain the general Landau-Lifshits formula (I will call it that, though I am not sure whether they are its original authors)

$$\Delta\theta = \frac{\partial}{\partial L}\left[\frac{2m}{L}\int_0^{\pi} r^2 \Delta U \, d\theta\right] \tag{6}$$

In particular, if we substitute (5) and (3) with $n=2$, 3, 4, we find

$$\Delta\theta = \begin{cases} -\dfrac{2\pi C_2}{Kp} & (n=2) \\ -\dfrac{6\pi C_3}{Kp^2} & (n=3) \\ -\dfrac{10\pi C_4(1+\frac{1}{2}\epsilon^2)}{Kp^3} & (n=4) \end{cases} \tag{7}$$

corresponding to perturbation potentials proportional to the inverse second, third, and fourth powers of r, respectively.

This establishes the tool we shall use in a moment; next, let us check how far the present theory is from Gerber's starting point (2). The modified Newton Law taking into account delays is given by (19), Sec. 1.8; the delay factor in the denominator of that expression must be incorporated in the potential as given by (13), Sec. 1.6. Moreover, the force in the radial direction is weakened by the cosine of the aberration angle, i.e. $\cos\beta \approx \sqrt{(1-\beta^2)}$. (We take all these factors as constant when we integrate force to obtain the potential.) Hence our expression for the potential is

$$U = -\frac{K}{r(1-\dot{r}/c)}\sqrt{\frac{1-\beta^2}{1-\beta^2\sin^2\theta}} \tag{8}$$

where θ is the angle between the radius vector $\mathbf{r}$ (centered in the sun) and the velocity of the planet. Thus (8) differs from Gerber's starting point by the factor of the square root, which we will now evaluate. By elementary differential geometry based on the elliptical orbit (15), Sec. 3.1, one finds

$$\cos^2\theta = \frac{\epsilon^2\sin^2\varphi}{1-2\epsilon\cos\phi+\epsilon^2}$$

where φ is the polar angle (ωt); hence the square root in (8), averaged over the orbit (φ from 0 to 2π), to second order in β and ϵ is

$$<1-\tfrac{1}{2}\beta^2\cos^2> = 1-\tfrac{1}{4}\beta^2\epsilon^2$$

For Mercury, with $\beta=1.597\times10^{-4}$ and $\epsilon=0.2056$, the square root is therefore not only very close to one, but since the coefficient of β^2 in the equation above amounts to only 1.06%, it is much closer to one than the delay parenthesis in the denominator. Our potential (8), therefore, is for all practical purposes equivalent to Gerber's starting point (2), to which we now return.

He derives the equation of motion via the Lagrange function Λ (kinetic energy T less potential energy U), and via the Lagrangian general equation of motion, which is explained in any textbook of theoretical mechanics (e.g. [Landau and Lifshits 1965], Chapt. I, Sec. 5),

$$\frac{d}{dt}\frac{\partial \Lambda}{\partial v} = \frac{\partial \Lambda}{\partial r} \tag{9}$$

Applying (9) and using (2), Gerber finds the acceleration of one body in the retarded gravitational potential of another:

$$\frac{1}{m}\frac{dT}{dr} - \frac{1}{m}\frac{d}{dt}\frac{dT}{d\dot{r}} = \frac{dU}{dr} - \frac{d}{dt}\frac{dV}{dr} = -\frac{\Gamma M}{r^2}\phi \tag{10}$$

where

$$\phi = 1 - \frac{3}{c^2}\dot{r}^2 + \frac{6r}{c^2}\ddot{r} \tag{11}$$

so that his equation of motion is equivalent to

$$m(\ddot{r} - r\dot{\theta}^2) = -\frac{K}{r^2}(1 - \phi) \tag{12}$$

which differs from the Newtonian IAAD equation (Sec. 2.1) by the appearance of the term containing ϕ. This is where we will leave Gerber; integrating his equation (12) will yield the corresponding potential, and the advance of the perihelion will then follow immediately from the Landau-Lifshits formula.

We will first show that the second term in (11) has no effect. It is found by expressing it in terms of r and substituting in (12). [This is done by using (5) and its derivative $r'(\theta)$, eliminating $\sin\theta$ by using $\cos\theta$ found from (5), and finally using $\dot{r} = r'(\theta)\dot{\theta} = r'(\theta)\,(L/mr^2)$.] The result is easily integrated to yield the perturbation in potential due to the second term in (11), i.e., due to the radial velocity of the planet:

$$\Delta U(\dot{r}) = -\frac{3KL^2}{mc^2}\left(-\frac{1-\epsilon^2}{r} + \frac{p}{r^2} - \frac{p^2}{3r^3}\right) \tag{13}$$

The first term in the bracket is proportional to $1/r$ and is equivalent to altering the value of the constant K (and only slightly at that), but it does not change the $1/r$ dependence of a Newtonian IAAD potential which leads to a simply closed conic. The other two terms, as can be seen from the first two lines of (7), exactly cancel, so that the effect of the second term in (11) on turning the orbit is zero.

Thus the only significant modification of the potential comes from the third term of (11), which depends on the radial acceleration. The product $r\ddot{r}$, if we multiply it by m, is force times radial distance, i.e. increase in potential energy, which must equal a decrease in kinetic energy:

$$\frac{6r\ddot{r}}{c^2} = -\frac{6(\frac{1}{2}v^2)}{c^2} = -3\beta^2 = -\frac{3L^2}{m^2r^2c^2} \tag{14}$$

Substituting this in the right side of the force (12) with its irrelevant velocity term discarded, and integrating over r, we obtain the potential in the form

$$U = -\frac{K}{r}\left(1 + \frac{L^2}{m^2c^2r^2}\right) = -\frac{K}{r} - \frac{K^2p}{mc^2r^3} \tag{15}$$

where we have substituted p from (16), Sec. 3.1, and this is equivalent to Einstein's approximation to the gravitational potential in curved space [1915].

From here on, the Landau-Lifshits formula would have saved both Gerber and Einstein several pages of calculations: from (15) and the second line of (7), we immediately have

$$\Delta\theta = \frac{6\pi K}{mc^2p} \tag{16}$$

It remains to substitute for p, which is also the semilatus rectum of an ellipse, i.e., $p = a(1-\epsilon^2)$, and since by our definition $K = \Gamma Mm$, we have

$$\Delta\theta = \frac{6\pi\Gamma M}{ac^2(1-\epsilon^2)} \tag{17}$$

which is not just similar or approximately equal to, but perfectly identical with (1). Chalk up another "Einstein minus zero."

There is, however, at least one important difference between Gerber's and Einstein's derivations. Though both arrive at similar gravitational potentials, Gerber gets there by classical physics based on hard physical concepts; Einstein arrives there after the abstract acrobatics of curved space-time in the intangible Temple of Tensors.

My streamlining of the derivation by the use of the Landau-Lifshits formula takes place only after the gravitational potential has been derived, and though it radically shortens the calculations, it has nothing to do with the essence of the underlying derivation in either case. What is important is that once more the Einsteinian equivalence has been shown due to the same mechanism: the propagation of force with a finite velocity.[1]

[1] Gerber actually worked with the propagation of potential or forced state (*Zwangszustand*); he claimed that propagation of force "repeatedly involves one in contradictions," but did not elaborate. I do not believe that the difference is critical; personally I regard force as a primary natural phenomenon, and potential as a man-made aid to simplify its calculation.

3.3. The Titius Series

In 1766, a German translation of a French book on astronomy (Charles Bonnet's *Contemplation de la Nature*, Amsterdam 1764) was published. The translator was Johann Daniel Titius (1729-96), professor of natural science at the University of Wittenberg, and pp. 7-8 contained a 22-line paragraph that was absent from the French original, a sort of translator's non-footnote inserted by Titius into the text. (The second edition of 1772 did show it as a regular translator's footnote under the main text.) The paragraph notes that the (mean) distances of the planets from the sun follow a certain law, and that where it shows a gap (between Mars and Jupiter, at the time) there must be another body to be discovered. The law amounted to

$$r_n = 4 + 3 \times 2^n \tag{1}$$

where r_n is the mean distance which is 4 units for Mercury (i.e. $n = -\infty$), and n equals 0, 1, 2, ... for successive planets. It was known as the "Bode" series in the last century and is now known as the "Titius-Bode" series, but since the German astronomer Bode (1747-1826) had nothing to do with it except popularizing it, and even then without giving the true discoverer credit, I see no reason to keep his name attached to the law.

In 1781 the Titius Law obtained a strong boost with the discovery of Uranus, which fit $n=6$ quite closely, and stimulated a search in the gap between $n=1$ and $n=3$, which Titius had attributed to an undiscovered satellite of Mars, but where in fact the planetoids of the asteroid belt were discovered.

However, when Neptune and Pluto were discovered, the agreement was less impressive, and in 1913, an American astronomer, Miss Mary Adela Blagg (1858-1944), published a new formulation of the law in the form

$$r_n = A(1.7275)^n [B + f(\alpha + n\beta)] \tag{2}$$

where A, B, α, β are constants given and discussed in [Nieto 1972]; the law essentially states a geometric progression, to which the factor in the square brackets represents a correction. The agreement is shown in the figure, also based on [Nieto 1972], in which the dashed straight line, representing the uncorrected geometric progression, has a slope corresponding to 1.7275. Other formulations have been attempted since then, but they differ only in the methods of expressing the correction to the straight line in the figure; what they have in common is 1) the geometric progression, 2) the constant ratio 1.73.

The regularity of the Titius series rules out coincidence; but what is even more striking is that the other three planetary systems against which (2) can be checked, namely the moons of Jupiter, Saturn and Uranus, not only again exhibit a geometric progression, but the same common ratio 1.73.

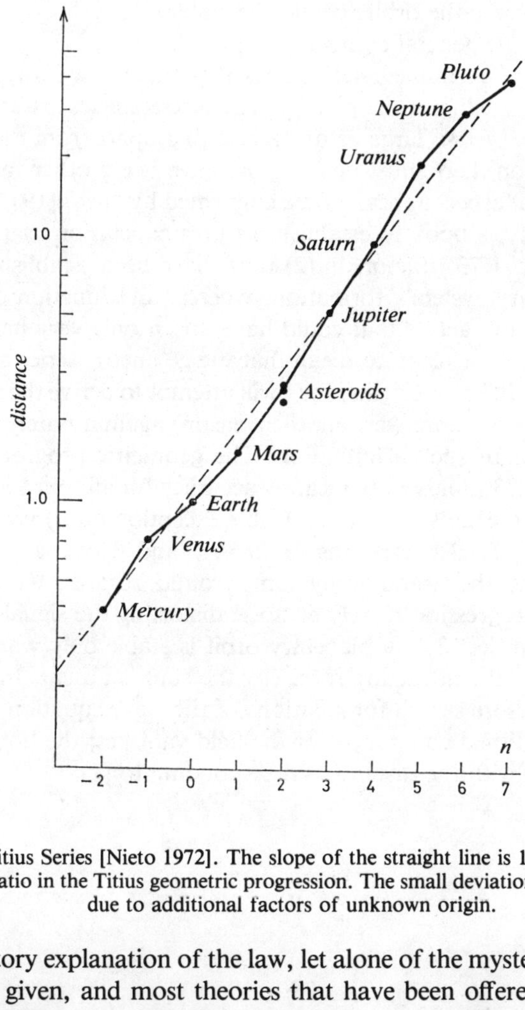

The Titius Series [Nieto 1972]. The slope of the straight line is 1.73, the constant ratio in the Titius geometric progression. The small deviations from it are due to additional factors of unknown origin.

No satisfactory explanation of the law, let alone of the mysterious number 1.73, has ever been given, and most theories that have been offered search for an explanation in the field of cosmogony, i.e. in the way in which these systems were originally formed.

But if the explanation lies solely in the original formation of the systems, then it seems strange that they should have preserved their regularity through the billions of years of their existence. Did no planet or its moon ever collide with a comet, an asteroid or other heavenly body? And if it did, what pulled it back to its "assigned" orbit?

In the following I will suggest that part of the explanation may be dynamic rather than cosmogonic. As in the electronic analogy, the aberrational component of the gravitational field in the modified Newton Law must somehow be balanced or eliminated; otherwise the orbits cannot be stable.

However, first some general coments.

A glance at the figure immediately shows that the Titius Law does not completely describe the spacing of the planets. The discrepancies are small enough to rule out coincidence, yet so large as to suggest that apart from the fundamental geometric progression there must be (or must have been) other factors at work. This purely technical aspect appears to be confirmed by Nieto [1972] who devotes an entire chapter of his book to establish his firm conclusion that the geometric progression, i.e. the 1.73^n factor, in (2) must have been established in an early period of a planetary system's formation, whereas the function defined by the bracket was caused by factors that could have arisen only very much later.

I will take Nieto's conclusion to mean that the geometric series and its "correction" are of entirely different origins, for I will attempt to derive the geometric progression only. There is, moreover, another reason, again a purely technical one, that speaks in favor of such a limitation. The geometric progression, and particularly the value 1.73, holds with uncanny accuracy for all four known planetary systems; the other constants (with the possible exception of α) vary quite wildly: A over a range of 7 : 1, B over a range of 285 : 1, and β over a range of 4.5 : 1. I will therefore, to use the jargon of my home ground, regard the factors multiplying the geometric progression merely as noise distorting the signal.

As pointed out in Sec. 3.1, a planetary orbit is stable only where the aberrational component of the attracting force (by the sun) is absent. In the following section we will therefore search for solutions of the wave equation (11), Sec. 3.1, in which the aberrational component of the field vanishes; the hope is that these zeroes will coincide with the distances corresponding to (2).

3.4. The Stable Planetary Orbits

We now return to the wave equation (11) of Sec. 3.1:

$$\nabla^2 \mathbf{F}_\theta(r,\theta) - \frac{1}{c^2}\frac{\partial^2 \mathbf{F}_\theta(r,\theta)}{\partial t^2} = \mu\ {}^0J_\theta\ {}^0M \tag{1}$$

where ${}^0M$ is the mass of the attracting body and ${}^0\mathbf{J}$ is non-zero only inside the planet (or at the point of the planet if we shrink it by means of the Divergence Theorem), a matter to which we will return presently.

The presence of the aberrational component F_θ of the force field would make the orbits of the planets unstable just as in the case of electrons, where the aberrational component was counterbalanced by the magnetic and self-induced electric field. In the case of the planets, however, for the reasons given in Sec. 3.1, we must search for a solution of (1) in which the aberrational component F_θ vanishes; and we shall find that it does so at discrete values of the distance r from the center of attraction – namely in a geometric series like the one given by the Titius Law.

Since the Titius Law applies to the *average* distances r, we shall in the following neglect the eccentricities of the planetary orbits and treat them as if they were circular. Except for Mercury (0.20) and Pluto (0.24), these eccentricities are quite small (under 0.06); but more important, there is no point in refinements for a law that has such significant discrepancies due to factors of other origin. The following will therefore be restricted to deriving the geometric progression that forms the essence of the Titius Law.

We eliminate time by letting the angular coordinate φ rotate with the planet:

$$\phi = \theta - \omega t, \qquad \frac{\partial^2}{\partial t^2} = \omega^2\frac{\partial^2}{\partial \phi^2} \tag{2}$$

Setting $\omega/c = \varkappa$, we obtain the following equation for the aberrational component of the field:

$$\nabla^2 F_\phi - \kappa^2\frac{\partial^2 F_\phi}{\partial \phi^2} = \mu\ \ {}^0J0 \tag{3}$$

We will first solve the reduced (homogeneous) equation; the fundamental system of two independent solutions will then be supplemented by the particular integral corresponding to the right side in (3). The reduced equation can be solved by separation of variables, which assumes that the unknown function can be represented as the product of two functions, each of which depends only on one of two independent variables:

$$F_\phi(r,\theta) = R(r)\Phi(\varphi) \tag{4}$$

Expressing the Laplacian in polar coordinates and dividing by $R\Phi$, we have

$$\frac{1}{r^2}\left(\frac{r^2R''+2rR'}{R}+\frac{\Phi''}{\Phi}\right)-\frac{\kappa^2\Phi''}{\Phi}=0 \tag{5}$$

When the value of φ is changed, it has no effect on (5), since r is by assumption an independent variable. This is possible only if Φ''/Φ equals a constant; and here we leave the well-trodden path.

In the case of a circular waveguide or a kettledrum membrane, which also lead to (5), this separation constant is always chosen as the negative square of an integer, so that the resulting sine-cosine solution for the Φ function in (4) is single-valued, i.e., it returns to the original value when φ is increased by 2π in going round the circle.

In our case, however, a sine-cosine solution makes no sense: the aberrational component (4) is strictly tied to the velocity of the planet, and therefore vanishes identically everywhere except for $\varphi=0$, corresponding to the location of the planet (which we will shrink to a point by the Divergence Theorem). Therefore the Φ equation involves a separation "constant" equal to $\pm S^2$ (we leave the sign open for a moment) only for $\varphi=0$ and equal to zero everywhere else; using the delta (impulse) function, the Φ equation is therefore

$$\Phi=\pm S^2\sum_{k=0}^{\infty}\delta(\varphi-2k\pi) \tag{6}$$

(We could equally well omit the sum and put up a barrier, as in contour integration, at $\varphi=\pm\pi$, or almost anywhere else, defining Φ in only one single-valued branch of length 2π. The value of Φ to either side of the barrier is zero.)

Thus, the aberrational force (4) now exists only for $\varphi=0$, as it should; however, its dependence on r in that direction is a continuous function $R(r)$. The periodicity and single-valuedness of the Φ solution is now ensured by the periodicity of the delta-function (k is an integer), which leaves us free to choose the sign of the separation constant S as we please.

Breaking with tradition, but violating no rules in the rationale for separating variables, I choose the separation constant positive: this will result in a positive delta function for Φ, which is neither surprising nor interesting, but the equation for R is now

$$r^2R''+2rR'-(\kappa^2r^2-S^2)R=0 \tag{7}$$

which is a generalized Bessel equation with the "wrong" sign before the parenthesis — the solution is a Bessel function of imaginary order. The reader who wishes to pursue this exact solution will find it quite tractable, for the function $J_p(z)/z^p$ (where J is a Bessel function) is an entire function for any complex p in the complex z plane, and therefore the only singularity of $J_p(z)$ is that of z^p, namely a branch point at $z=0$ when p is not a real integer.

However, using Bessel functions on this equation is shooting a sparrow with a cannon, for it will be seen that the first term in the parenthesis equals β^2, which even for Mercury, the fastest planet, is of the order of 10^{-8}, and therefore negligible compared with one. On discarding it in (7), we obtain the much simpler Euler-type equation

$$r^2R'' + 2rR' + S^2R = 0 \tag{8}$$

which is solved by setting $R = r^u$ and solving for u after substitution in (8), yielding the two complex conjugate roots

$$u = -\tfrac{1}{2} \pm i\sqrt{S^2 - \tfrac{1}{4}} = -\tfrac{1}{2} \pm is \tag{9}$$

where s denotes the imaginary part.

Thus the solution of the reduced equation (in the $\varphi = 0$ direction) is

$$R = r^{-\frac{1}{2}}(Ar^{is} + Br^{-is}) \tag{10}$$

It remains to find the particular integral of the full equation (8) with the right side equal to

$$\mu\ {}^0J\ {}^0M = \mu\ {}^0Mmv = \mu\ {}^0Mmp\sqrt{KM}r^{-\frac{1}{2}} \tag{11}$$

where m is the (point) mass of the planet (in coulomb-equivalents or kilograms with or without superscript, respectively), and where we have substituted for v from (12), Sec. 3.1 (with $r = a$). This can be done by Lagrange's method of varying the constants, or more quickly by a guess at the type of solution ($C/\sqrt{r}$), substituting it in the full equation (8), and if successful, determining the constant [$C = {}^0M\mu\sqrt{(Km/s)}$]. This yields the general solution of the unreduced equation (8):

$$R(r) = r^{-\frac{1}{2}}\left(r^{is} + Br^{-is} + \frac{{}^0M\mu\sqrt{Km}}{s}\right) \tag{12}$$

Since

$$r^{is} = \exp(is\ln r) = \cos(s\ln r) + i\sin(s\ln r) \tag{13}$$

it follows that the general solution (with A and B chosen as complex conjugates of each other to make the solution real) is

$$R(r) = r^{-\frac{1}{2}}\left[C_1\cos(s\ln r) + C_2\sin(s\ln r) + \frac{{}^0M\mu\sqrt{Km}}{s}\right] = 0 \tag{14}$$

As the range of r is truly astronomic, the argument of the trigonometric functions will go through several points where (14) vanishes. Let $r = \varrho$ be such a point; then (14) will also vanish whenever

$$s\ln r = \rho + 2k\pi \tag{15}$$

where k is an integer. The function $R(r)$, and with it the destabilizing component (4), will therefore vanish at the points

$$r_k = \exp\left(\frac{\rho + 2\pi k}{s}\right) \tag{16}$$

Hence, if we denote

$$r_0 = \exp\left(\frac{\rho}{s}\right), \qquad b = \exp\left(\frac{2k\pi}{s}\right) \tag{17}$$

the planetary orbits will be stable at the discrete distances

$$r_k = r_0 b^k \tag{18}$$

which is the Titius Law: once again field-referred velocities, without any additional assumptions, have yielded a result that the Einstein theory has not supplied.

I will therefore stake my claim to "Einstein Plus Two" for this item, but only in a very subdued manner, for it differs sadly from the "Einstein Plus One" item of discretely stable electron orbits. The latter were verified, for example, by deriving Planck's constant by electromagnetic theory alone (Sec. 2.9), whereas in the case of the Titius Law, I know of no way to verify my derivation by an independent method, nor to determine the constant s (let alone r_0) numerically.

There is, for example, the constant b, which Miss Blagg [1913] found equal to 1.7275 for the best fit in the solar system. For Jupiter's satellite system, she ended up with the same mysterious number 1.7275, and this number held again for the remaining two known satellite systems, those of Saturn and Uranus. To some extent the value of this constant depends on the type of formula with which one approximates the none too precise Titius Law; however, the other method by Richardson (see [Nieto 1972]) yields a constant that agrees with Blagg's to three significant figures, so that we can accept

$$b = 1.73\ldots \tag{19}$$

It would be nice to come up with just this value in (17) by somehow determining s, related to the separation constant S^2 by (9). If, for example, s had to be an integer plus half (as is the case for the corresponding Bessel equation of *real* order), $s = 11.5$ would do just right, for

$$\exp(2\pi/11.5) = 1.7269\ldots \tag{20}$$

but even then the odds against this being accidental would be less than overwhelming.

The only point that provides some qualitative support to the derived result is Blagg's constant b: as given by (17), it does not depend on a particular planet, or even on a particular planetary system, and this is in agreement with observation of the four known planetary systems. To the contrary, the constant r_0 is determined by both the planetary system through m and K in (14), and this is again confirmed by observation – it varies over a range of roughly 7 : 1.

For the rest, I must content myself with having derived the functional form of the Titius Law (18), but without being able to derive the numerical values of the constants. Although this is comparable to other theoretical laws with experimentally measured constants, I will readily concede that "Einstein Plus Two" stands on far shakier ground than "Einstein plus One."

3.5. Siblings, Twins, or One Identical Child?

We have found it possible to derive all of Einstein's results, plus two more, by assuming that forces propagate from their source to the objects they act on with velocity c, the velocity of light, with respect to the local force field. The fact that this hypothesis leads to agreement with observation for both electrons and planets, that is, regardless of whether the field propagating from its source is electric or gravitational, suggests that electric and gravitational forces might not be siblings, but the same phenomenon. This suggestion gets some support from the macroscopic magnetic field, which is proportional to the velocity of a charge with respect to neutral matter (Sec. 1.4).

Since all macroscopic matter (the only type where gravitation has been observed) is known to consist of positive nuclei and and negatively charged electrons, the idea that gravitation is ultimately due to electric forces appears plausible; it has been proposed by several scientists, of which Walther Ritz [1908, 1911] may not have been the first.

Let us see what stands in the way of such a proposition:

1) there are two types of electric charges, positive and negative, but there is only one type of mass;
2) there is no charge without mass, but there is mass without charge;
3) mass has inertia, charge does not;
4) mass varies with velocity, whereas charge is an invariant;
5) electric force can be screened, but nothing is opaque to gravity;
6) we have always been told the two are different.

Of these, only no. 6 is a serious obstacle; the others may be false, removable, or of our own making.

No. 3, for example, is false: inertia is the reaction to a force, the resistance to being accelerated. The phenomenon is observed with charges, or at least with their fields, just as surely as with masses: see Secs. 1.7 and 2.6. When a large number of charges – a current – is accelerated, we call the phenomenon "self-inductance;" when a single (point) charge is accelerated, the same resistance to acceleration sets in, but we have not been paying much attention to it in our textbooks. The classics of the late 19th century worked with the concepts of "electromagnetic mass" and "electromagnetic momentum;" the concepts have gone out of fashion, but they have not been refuted. In both cases, the resistance to acceleration is proportional to the acceleration, and to the mass or charge of the object that is resisting. "Proportional to" is made by nature; "equal to," when it depends on units, is made by man. Any difference between the inertia of mass and the inertia of charge is therefore of our own making.

No. 4 is clearly of our own making, too. The correct statement from the point of view of the present theory is that inertial mass is velocity-dependent, whereas gravitational mass, like its analog electric charge, is invariant — or we can make it so, if we choose to regard velocity-dependent field and aberration effects as separate factors rather than incorporate them in the value of gravitational mass or electric charge.

No. 2 becomes at best trivial in view of these two realignments of concepts (gravitational mass the analog of electric charge and inertial reaction the equivalent of self-inductance). The nearest valid statement to no. 2 is then that neither gravitational mass nor electric charge is without inertial reaction, with which no one (even an Einsteinian, if he knows his electromagnetics) should have any quarrel. However, as an objection to the identity of gravitational and electric force, no. 2, as it stands, becomes an exhibit of flawed logic. Suppose one could overcome the experimental difficulties — say by charging and discharging the leaden balls in the Cavendish torsion scales. No one doubts that the balls would move; but how does that decide whether a force of an entirely different character was activated or whether the gravitational force of the leaden block was made to change its value?

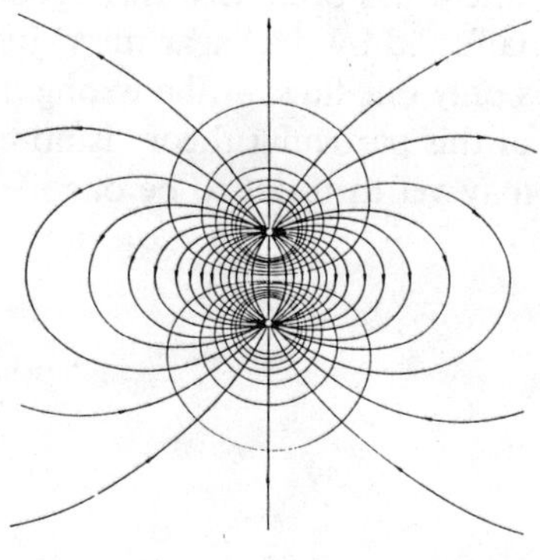

The familiar field of an electrostatic dipole includes the straight line closed via infinity.

That leaves no. 1, which says that an electric force may either attract or repel, but a gravitational force can only attract. Now we all know that positive and negative charges neutralize each other — but do they? Consider the field of an electric dipole, shown above, including the (straight) line of force closed via infinity. A distant dipole, also supposedly neutral at large distances, has such a line, too, and the two dipoles will by all rules of geometry switch their straight line of force from infinity to each other. Depending on their orientation, they will attract or repel each other; but the repulsive orientation is unstable, so in essence these two allegedly neutral configurations will end up attracting each other.

I hope I have made no quick converts to this theory, for I am about to bury it.

The force between two electric dipoles varies as the inverse *cube* of the distance between them, as is easily shown from the field of one dipole derived in virtually all textbooks. The same inverse cube would apply to two "masses" of dipoles. For multipoles the inverse power of the distance is even higher [Stratton 1941]. Gravity, however, acts by the inverse *square* law just like single electric charges, which appears to make the theory untenable.

The reason why I have nevertheless mentioned it is to show what a comparatively near miss it is. *It fails only by the wrong dependence on distance.* However, this failed theory is not needed to remove the other objections, which are removed by no more than a conceptual realignment: charge and gravitational mass are twins not just because they appear in the same place in the Newton-Coulomb Law, but because both exhibit the same type of inertial reaction when accelerated, the measure of resistance to acceleration being given by electromagnetic and Newtonian inertial mass, respectively.

Such a conceptual realignment from contemporary thinking, beyond leading to agreement with experiment without perverting space and time, will also convert electromagnetism and gravity from siblings to twins.

This conceptual realignment has been used throughout this book (see table on opposite page), and is unaffected by the "near miss" just discussed.

However, since there is only one flaw — the wrong dependence on distance — to overcome, the failure of this particular theory is no final refutation of the idea that the supposed twins may yet turn out to be one identical child.

Invariance...

Quantity	*Old*	*New*
space	velocity-dependent with respect to observer	invariant
time	velocity-dependent with respect to observer	invariant
inertial mass	velocity-dependent with respect to observer	velocity-dependent with respect to field
electric charge	invariant	invariant
gravitational mass	velocity-dependent with respect to observer	invariant

...and Affinity

Newtonian inertial mass	total inertial mass is equivalent to gravitational mass	analogous to electromagnetic mass
gravitational mass	equivalent to total inertial mass	analogous to electric charge
electromagnetic mass	not disputed, but ignored (lumped in with Newtonian inertial mass)	analogous to Newtonian inertial mass

3.6. Inertia

Now let us turn to a fundamental, yet still not properly understood phenomenon: inertia.

Newton's discovery of the inertia principle is surely one of the greatest discoveries in the history of mankind; yet the inherent nature of inertia remains unexplained.

Some classics, and more recently Barnes [1983], tried to explain its origin by electromagnetic forces. The present theory (Sec. 2.6) asserts that the inertia of charged particles is in part due to their electromagnetic field, but that part becomes negligible for macroscopic bodies as the ratio of charge to dimension decreases. These attempts may provide some insight, but no full answer, for in essence they only shift the puzzle, partly or fully, from general physics to electromagnetism.

For the inertia of *uncharged* matter, the present theory provides no explanation, though it does not contradict the observed results.

If we use the method of Sec. 2.7, assuming nothing but the gravitational inverse-square law and the propagation of force and potential, we find by repeating the procedure with $-\Gamma M^2$ substituted for $q^2/(4\pi\epsilon)$

$$\mathbf{F} = -\frac{M^2\Gamma}{c^2 R}\,\dot{\mathbf{v}} \tag{1}$$

where $\mathbf{F}$ is the force exerted by the asymmetrical gravitational field on the accelerating mass that created it, and R is the radius of a sphere with gravitational mass M. Hence the inertial mass *due to this mechanism* is

$$m' = -\frac{M^2\Gamma}{c^2 R} = -6 \times 10^{-27}\,\frac{M^2}{R} \quad \text{kg} \tag{2}$$

The additional contribution of this mass to the inertial mass as known from experiment at low velocities,

$$m = M \tag{3}$$

is utterly insignificant, for if we consider (2) a correction to (3), we have

$$m = M\left(1 - 6 \times 10^{-27}\,\frac{M}{R}\right) \quad \text{kg} \tag{4}$$

and the second term will be seen unmeasurably small in the world from the electron (M/R of order 10^{-16} to the earth (10^{-18} and sun (10^{-6}).

In short, the effect of the gravitational field on accelerated mass is insignificant; where the vastly larger observed inertial reaction to a force comes from remains an unanswered question.

To my knowledge there have been only two significant attempts to explain the actual origin and nature of inertia, that is, to throw light on the mechanism that makes an uncharged body resist acceleration.

The two efforts known to me were made by Walther Ritz and Albert Einstein.

Ritz [1908, 1911] pointed out that *les particules entièrement hypothétiques* ejected by material bodies (Sec. 1.3) would have a radially symmetrical density with respect to the emitting body in any system in which the body moves with uniform velocity, but as soon as the body accelerates, the density of the particles would increase in front of the body and decrease behind it as the emitting source catches up with, or draws away from, the particles it has ejected (see figure, p. 114). Ritz calculated the inertial force (as well as the electric and gravitational force) from the assumption that the force is proportional to the density of the particles. As discussed in Sec. 1.3, Ritz's ballistic assumption is untenable for the propagation of light from macroscopic moving sources, though its alleged refutation by double stars is invalid. However, the propagation of light is not necessarily related to inertia, and for the latter case, I know of no evidence either to support or to contradict Ritz's theory, so that I can have no opinion on it. But the idea is not in itself absurd: Ritz's theory, which is a genuine relativity theory, says that when an aircraft accelerates along the runway during take-off, I am being pressed to the back of my seat by an imbalance between something before and behind me, something that has its source right there in the matter that is subject to inertial reaction.

True or false, I see nothing absurd in the idea as such.

That is not the case with Einstein's idea that inertia and gravitation are identical. The idea may be logically flawed, as recently pointed out by Brown [1983], but there is something else that I find hard to accept. To make the two numerically equal, one needs a lot of matter, more than there is in the solar system, and Einstein had to invoke all of the universe (in crooked space, bent back onto itself) to make the two forces identical. I do not doubt that mathematically this works out beautifully as his equivalences always do; but physically it means that what presses me to the back of the seat on the runway is the attraction by distant galaxies that may no longer be there.

As a physical reality, I find this fantastic beyond belief.

Epilogue

Epilogue

As I said in the Preface, it is probably too much to hope that this attempt to provide an alternative to the Einstein theory will succeed at first try, and I have braced myself for the discovery of serious errors (let alone misprints).

But at the very least it is my hope that this essay will stimulate others to base physics on the primitive concepts of space and time as they have naively been understood since Euclid and the other Alexandrians founded science two and a half millenia ago; to base it on the genuine Principle of Relativity, undistorted by twisting space and time; to discern between truth and equivalence, though both may yield the observed result; and above all, to strive for insight rather than mere description.

I harbor this minimum hope especially for the occasional young engineer or physicist who goes to college thirsting for real knowledge rather than just for a degree, and who is smart enough to realize that the computer is no more than a fast and powerful tool that will produce stacks of printouts, but no genuine insight. He is handicapped by lack of knowledge and experience, but has the advantage of a mind not yet molded by conventional wisdom.

But of course, I have maximum hopes, too; and should I have been fortunate enough to be on the right track, then paraphrasing the words of the greatest genius of all times, I will say: If I have seen further than others, it is because I peeked past the giants who were blocking the light.

APPENDIX

The Devil's Advocates

This is the third version of the book, reworked from scratch. Some or all versions were reviewed by four staunch Einsteinians to whom I had submitted the manuscript for criticism and comment:

Howard C. Hayden, Professor of Physics at the University of Connecticut, Storrs;

Homer G. Ellis, Professor of Mathematics at the University of Colorado, Boulder;

and two others who did not wish to have their names associated with the book:

Gene, a physicist with a private company in Palo Alto, Calif., and

Paul, a physicist with the Lawrence Livermore National Laboratory.

I am most grateful to all four for having pointed to numerous errors (and bloopers that I would rather not talk about), and for sending me back to the drawing board for inconsistencies that I had overlooked. Paul, for example, found a discrepancy of a factor of 2, which I at first thought an arithmetic error, but which in fact not only resulted in a full year of additional work, but destroyed my illusion that I had derived the numerical value of Planck's constant in terms of independently measurable constants.

I am particularly thankful to Prof. Howard C. Hayden, who spent more time than anyone else in discovering errors and inconsistencies and raising objections to my presentation. I hope it is superfluous to say that he is in no way responsible for the errors and inconsistencies that doubtlessly remain, and neither are the other reviewers. Such errors are exclusively my responsibility; there would have been many more without these four "devil's advocates," especially Prof. Hayden.

The great majority of objections have become pointless since I reworked the book. The corresponding errors have been eliminated, or they have become moot by developing the theory differently. But a number of points remain where I cannot agree, and this appendix is devoted to the discussion of the more important ones.

Time dilation and sacrifices

"I would like to see a somewhat different summary than is contained in the table on p. 187," writes Prof. Hayden [who sets $\gamma = \sqrt{(1-\beta^2)}$]. "To explain the rather odd dependence of velocity (such as $m\mathbf{v}\gamma$ for momentum) one is required to *sacrifice* some conventional notions. Einstein sacrificed space and time in order to preserve the speed of light and covariance. Experimental 'justifications' have shown that particles (e.g., muons) keep the same time in their own coordinate system. Masses, in the sense that all scalars must be covariant, are *not* variant, but rather constant. (Again, it is a matter of determining where the γ goes.)

"Your model makes other sacrifices. Space and time are preserved, inertial mass is not. For muon results, you postulate that the processes giving rise to radiation are always slower by a factor of γ, similar to Ives-Stilwell results, which are derived, if my reading is correct. This may be an entirely reasonable assumption, but you should list it as one of the sacrifices. Electric charge is preserved, but its distribution is not. In Ives-Stilwell, the *effective* charge is reduced, but your table shows charge as constant. The matter should be clarified in the summary. . ."

First, let me get some minor matters out of the way.

Electric charge is preserved, but its distribution is not. Quite so, but this has little to do with Einstein's theory or mine; equation (11) on p. 58 follows exclusively from the Maxwell equations, and (16) on p. 61 is a straightforward integral. This makes charge constant in all theories (including ether) and charge density variable in all of them. Even in Einstein's theory, which gives charge density *formally* the same appearance due to length contraction, it is numerically different for observers in different inertial frames: a constant charge is compressed into a smaller volume for the observer who moves past it at a high enough velocity.

As another minor point, the effective charge on p. 81 (I called it "fictitious") is a just-as-if quantity used as an artifact for a shorthand calculation. It has no more physical significance than a virtual image or the "pretense" that the kinetic energy of a bullet has quadrupled because it has four times the previous mass, when in fact the mass is the same, but its velocity has doubled. As a physical reality, I take charge as invariant, that is, as constant regardless of its velocity, whatever rest standard it may be referred to.

But now to the serious objection. My basic assumption is that the velocities that matter, the velocities that will make the Maxwell equations valid, are the velocities with respect to the traversed field. Using this basic assumption, and postulating nothing otherwise, I *derive* the frequency emitted by the moving hydrogen ions in the Ives-Stilwell experiment and find it shifted as in (11), p. 81. For the case of radioactive decay, the timing mechanism is unknown (beyond the general conditions that result in the random individual decays being Poisson-distributed), so that I cannot perform an equally detailed calculation. However, for both ions and muons the energy is proportional to the frequency — to the average frequency of decays in the latter case. I therefore assumed — yes, this is an additional assumption, but not one peculiar to my theory — that their frequencies would also be shifted in the same way.

Now compare this with the Einsteinian approach. For both of us, the Ives-Stilwell experiment, like all other time dilation experiments, remains one without a control experiment: none of us has sat on a moving hydrogen ion or on a muon to measure its inherent frequency (of radiation or decay) in its own frame of reference. The Einsteinians' assumption of observer-dependent velocities tells them that the inherent frequency has not changed; mine about field-referred velocities tells me it has. For muons, neither of us knows the timing mechanism that produces the average frequency, or equivalently, the halflife. With an equal measure of arrogance (if any), we thus both transfer the results of our basic assumptions from the known case to the case with the unknown mechanism. So in both cases, as far as velocity dependence goes, we are quits.

But *in addition to* the assumption about which type of velocity makes the Maxwell equations come true, the Einsteinians have sacrificed space and time; I have sacrificed nothing. I lack their authority and prestige, but they are short of simplicity.

Field rules

"Especially, you should point out (in the summary)," continues Prof. Hayden, "just where the unconventional assumption of referring velocities to the field occurs. Remember that in Einsteinian methods, one can place the coordinate system in any inertial frame. In the case of an electron orbiting at $c/137$, the field of the proton is fixed, but that of the electron moves rapidly. Since velocities with respect to the locally dominant field are to be used, I take it to mean that the dominant field is the moving, accelerated field, conveniently anchored, let us say, halfway between the proton and the electron. Is it here that one departs from one of the reasonable Einsteinian coordinate systems? (Einstein would be happy using the proton's center, for example.) In the *Sacrifices, Einstein's and Mine* section, you could do a lot to clarify the differences.

"What about helium, which has two electrons? How does one anchor to the field? Assuming that the anchoring is the sacrifice to be made, do you have an algorithm for doing it, given the difficulties of determining the electron's locations? (Mind you, the QM method involves its own peculiarities, but at least by now the methods are well worked out.)"

I am as free as Einstein to choose my coordinate system, regardless of the dominant field. The rules for the field within those selected coordinates are given on p. 48: the dominant field is the one in the immediate neighborhood of the considered particle, and the irrotational part of the particle's own field (typically, its Coulomb field) can be ignored, since it must result in a net force of zero.

Thus, in the case of helium, both electrons are acted on by the traversed field, the field of the proton; and the proton is acted on by the fields of the electrons which sweep past it as explained on pp. 51-52. Such words as *traversed* and *sweep past* imply that I have anchored my coordinate system in the proton as Prof. Hayden suggests, but this is not at all mandatory. For the purely mechanical aspect of the planetary motion I have no quarrel with Einsteinians, and the most convenient place to anchor them is in the center of mass. But for the electromagnetic aspect I am freer than Einstein in choosing an origin, for the forces acting on the charges are utterly independent of my coordinates: they are determined by their field-referred velocities, which are equally large for any one of the two electrons and the proton (the field of each electron sweeps past the proton with the same velocity as the electron traverses the proton's field). This will result in a Biot-Savart force that exactly cancels the destabilizing contribution of the aberrational component (p. 71) – in the helium or any other atom.

For Einsteinians things are less simple. If they choose the proton center for an origin, they get no Biot-Savart force, if they choose the center of mass, they get a very tiny one in the opposite direction to mine. This force is not (yet) amenable to measurement; but in the Einsteinian conception its size and direction are variable and determined by the whim of the observer choosing his coordinates.

"Just where exactly the unconventional assumption of referring velocities with respect to the field occurs" – it does not formally occur in the Maxwell equations proper, but in the definition of current density $(\mathbf{J} = \varrho\mathbf{v})$, in the magnetic part of the Lorentz force $(q\mathbf{v} \times \mathbf{B})$, and in any space-time transformation brought in by an electric charge moving through an electromagnetic field. The velocity of an observer and his coordinates is irrelevant, as it leaves the relative, field-referred velocity of the charges unaffected.

Angular momentum and subzero levels

"In the matter of angular momentum, there is the Quantum Mechanics (QM) value of zero for the ground state of the hydrogen atom, but the Bohr value of 1 (units of $h/2\pi$) for the same state. Your book leads to Bohr on one hand, and to QM on the other. The QM definition of angular momentum is not the same as mvr, so the solution may lie therein."

Very probably it does, and for the following reason: the angular momentum of an orbiting electron is not directly measurable; it determines the energy, which again is not measurable absolutely, but only as a difference in levels. [The energy of the orbiting electron is by (11) and (21), Sec. 2.1, proportional to the inverse square of the angular momentum.] Mathematically speaking, the solutions of differential equations, whether of motion or by Schrödinger, may differ by an additive constant (which is what we have here); its value is determined by the initial conditions. The initial condition (18), p.121, most definitely uses mass, velocity, radius and hence angular momentum in the strictly classical sense; if that meaning is changed to some formalism, it is not surprising that the value of the additive constant changes also.

Apart from that, as Prof. Hayden notes elsewhere, I have derived only the "time-independent" form of the Schrödinger equation, which does *not* contradict the old Bohr model. The difficulty seems to stem from the "time-dependent" Schrödinger equation involving the Hamiltonian of the wave function and its first derivative with respect to time. I have not included Prof. Hayden's comments with regard to the derivation of *that* equation here, as they appear to be a suggestion for further research and for desirable results rather than an objection to the theory as it stands. In any case, for the issue at hand, the discrepancy is one between the Bohr theory and the time-dependent Schrödinger equation; there is none between the Bohr theory, my own theory, and the time-independent Schrödinger equation derived from it.

"Your hypothesis of standing velocity waves suffers one flaw of arbitrariness, and that is the question of why the electron does not just go into the nucleus, radiating as it goes. That is, why can it transit from the $n=5$ state to the $n=1$ state (in truth, your model does not provide the mechanism, lacking the $\partial/\partial t$ term), but not from the $n=1$ state into the nucleus?"

First, as a minor matter, I do not think the velocity wave should be classed as an "hypothesis;" it is derived in Secs. 2.2 through 2.4 without new assumptions. But to the essence of the question, as of now I have no good answer, for I have derived only the stable states where the energy balance does match, that is, the states corresponding to positive integers. The transients, associated with the spiraling from one stable orbit to another, could in principle be worked out by the same method, as I have, perhaps somewhat glibly, pointed out in the text. But since I have not actually performed these rather nasty calculations, it remains only a hope, not a conviction backed by derivation, that when the orbital energy is plotted as a continuous function of the radius from zero to infinity, it will have minima at the stable orbits corresponding to positive integers, the valley at $n=1$ being separated from the nucleus ($n=0$) by a high potential barrier. This is admittedly unverified expectation; it is part of the big green pasture waiting to be grazed.

Geometry of space and force

Prof. Homer G. Ellis, whose oral comments I reproduce here as best I remember them, objected to three points that one might well associate with the definition of space. First, he objects that lines of force and equipotentials do not result in a well-defined space. I cannot see why not: however naively, I regard their points of intersection as the analogy of the knots in a fisherman's net, which I use as a standard of rest. Intersections of lines of force and equipotentials can be measured, at least in principle, and if the resulting coordinate net is mathematically unclean, I have little doubt that mathematicians can formulate the same idea more rigorously.

In Sec. 1.4 and the figure on p. 51, the interaction between two moving charges is dependent on the transversal velocity in a polar coordinate system, yet everywhere else, in particular in Secs. 1.10.3 and 1.10.4, I talk about the transformation of velocity as "the simple difference" between the two coordinate systems, Prof. Ellis further objects.

But I believe that this is neither inconsistent nor unprecedented. When a quantity (here the force between the two charges) is not dependent on a velocity in direct proportion, but via some operation — such as the cross product involving velocity in angular momentum or magnetic force — then it is both logical and customary to transform the velocity in accordance with the given transformation first, and then to use the result for substitution in the operation. That is how it is done with the Galileian transformation for angular momentum or magnetic force, and that is how it is done here: an observer moving past the charges shown on p.51 with half the speed of light will determine the effect-producing velocity, which is the velocity of one charge in the field of the other, by first taking the simple (vector) difference between their individual velocities and his own. Then he will apply the method of p. 51, which remains unaffected by his own velocity.

The last of Prof. Ellis' objections in this category is that the wave equation (1) in Sec. 3.4 is invalid, for the wave equation derived from Maxwell's equations is valid only for fields, and I am using it for a force that is not Lorentz-invariant.

I will refrain from disputing the point on general grounds, for it is numerically insignificant. The wave equation is linear and remains valid if its variable $\mathbf{E}$ is multiplied by the constant q. If I pretend that $q\mathbf{E}$ stands for all of the force, then my error in neglecting its magnetic component is (for Mercury, the fastest planet) of the order 10^{-8}. Alas, my book contains worse approximations.

Observables vs. inferrables

The charge redistribution emerging from the Maxwell equations as in (11), Sec. 1.6, is real, I claim; but it is the charge distribution, not space itself, that contracts.

But then, asks Prof. Ellis, the crystals in your measuring rod will contract, too; and how are you going to measure lengths?

This is a matter of philosophy. For argument's sake I will concede (without any experimental evidence) that even neutral matter contracts at high velocities. When our measuring instruments deceive us, do we correct for their errors or do we give in to their false readings? I remember when the moving-coil galvanometer was the most accurate instrument for measuring very small currents. It was so sensitive that it could, at times, be affected by the earth's magnetic field. Did we proclaim Ohm's Law to be dependent on geomagnetic coordinates? No; we switched the leads to reverse the current and took the average of the two readings. (We should probably have reversed the position of the whole meter.) To this day currents are rarely measured directly; they are *inferred* from electronically measured voltages. And to this day we use imperfect measuring instruments to measure as perfectly as rational inference will permit.

Observables or inferrables? We are back to the railroad track in the Introduction.

Aberration

Gene saw only the first version of the book, and most of his objections have become moot — for example, the discrepancy of the electron radius, discussed on pp. 141-142. He also objected to my treatment of aberration, which I simply applied to propagating force as if it were a propagating wave or a cloud of moving lead shot. I still do so without apology, for I regard aberration as a geometrical phenomenon dependent on velocity only. The point has become moot, or at least muted, since aberration no longer plays the fundamental role it played in the first version. Nevertheless, I have evidently not made a good job of explaining the direction of the aberration of force, for Prof. Hayden found my figures and text confusing. Let me try again with an additional example here.

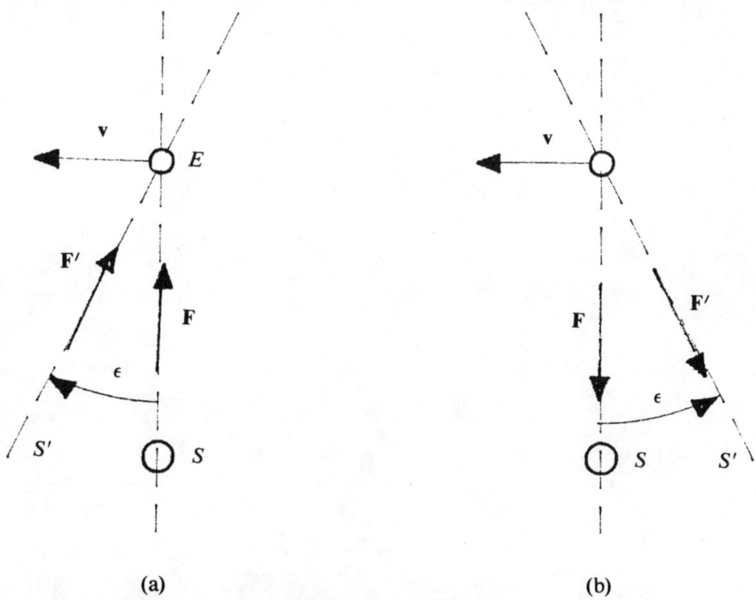

Aberration, (a) toward the velocity, (b) away from it.

Mr. E (representing the Earth) is running with velocity **v** and Mr. S (representing the sun emitting light) is throwing rocks at him as in figure (a). If the velocity of the rocks were infinite, they would hit Mr. E in the left shoulder along the true direction of their flight *SE.* But if Mr E's speed is commensurate with those of the rocks, they will hit him in the left part of his chest; since he is running into the rocks, he will perceive them coming from the aberrant direction *S′E,* which deviates from the true direction *SE* by an aberration angle ϵ *toward* the direction of the velocity.

In figure (b), Mr. S is not throwing the rocks, but sucking them in with a giant vacuum cleaner (representing the sun's gravitational attraction, or the attractive Coulomb force on a charge). This time Mr. E would get hit in the right shoulder if the velocity of the rocks were infinite; in fact he will get hit in the right part of his chest, for the rocks appear to come from the aberrant direction ES', which deviates from the true direction ES by an aberration angle ϵ *away from* the direction of the velocity.

That is what I evidently did not succeed in conveying in the small print on p. 31. In both cases, of course, the velocity component of the aberrant force is directed *against* the velocity of the runner, just as the transversal component of both sunlight and the sun's gravitational attraction is directed against the orbital velocity of the earth.

References

BARNES, T.G.,, "Electrical Explanation of Inertial Mass," *J. of Classical Physics* (Box 492, New York, NY 10185), vol. 2, Part II, July 1983. This is evidently a reprint of a paper with the same tile in *Creation Research Quarterly* (2100 Greenfield Dr., El Cajon, CA 92021), March 1983; parts of it are included in a wider-ranging book, *Physics of the Future*, Master Books (Box 15908, San Diego, CA 92115), 1983.

BECKER, R., *Electromagnetic Fields and Interactions*, Dover Publ., New York, 1964.

BECKMANN, P., "Equivalent electromagnetic constants of moving media," *Proc. IEEE*, vol. 58, pp.800-801 (1970).

BECKMANN, P., and P. MANDICS, "Test of the constancy of electromagnetic radiation in high vacuum," *Radio Science*, vol. 69-D, pp.623-628 (1965).

BOHR, NIELS, "On the constitution of atoms and molecules," *Philosophical Magazine*, vol. 26, pp.1-19 (1913).

BONCH-BRUYEVICH., M.A., and V.A. MOLCHANOV, "Novyi opticheskiy relativistskiy opyt," *Optika i spektroskop.*, vol. 1, pp.113-124 (1956).

BORN, M., *Atomic Physics*, 7th Ed., Engl. Transl., Blackie, London 1963.

BROWN, G. BURNISTON, *Retarded Action at a Distance*, Courtney Publications (95/115 Windmill Rd., Luton, Bedfordshire, England), 1982.

BRADLEY, JAMES, "Account of a new discovered motion of the Fix'd Stars," *Phil. Trans.*, vol. 35, p.637 (1728).

CHAMPION, F.C., *Proc. Roy. Soc.*, vol. A-136, p.630ff. (1932).

DE SITTER, W., "Ein astronomischer Beweis für die Konstanz der Lichtgeschwindigkeit," *Physik. Z.*, vol. 14, p.429 (1913).

DINGLE, HERBERT,, "The origin and present status of the special theory of relativity," -*Science Progress*, vol. 48, pp.201-219 (1960);*Science at the Crossroads*, Brian & O'Keefe, London 1972.

EINSTEIN, A., "Zur Elektrodynamik bewegter Körper," *Ann. d. Phys.*, vol. 17, p.891 ff. (1905a); "Ist die Trägheit eines Körpers von seinem Energiegehalt abhängig?" *Ann. d. Physik*, vol. 17, (1905b); "Erklärung der Perihelbewegung des Merkur aus der allgemeinen Relativitätstheorie," *Sitzungsberichte der preussischen Akad. d. Wiss.*, vol.47, no.2, pp.831-839 (1915).

ELLIOT, R.S., *Electromagnetics*, McGraw-Hill, New York 1966.

ESSEN, L., *The Special Theory of Relativity: A Critical Analysis*, Oxford Science Research Papers 5, Oxford University Press, Oxford 1971.

FISCHBACH, E., and ARONSON, "Reanalysis of the Eötvös experiment," *Phys. Rev. Letters*, vol. 56, no. 1, pp. 3-6, (6 Jan. 1986).

FISCHER, G.E., and Y. MURATA, "A beam monitor system for high-intensity photon beams in the multi-GeV range," *Nucl. Instr. and Meth.*, vol. 78, pp.25-39 (1970).

FRISCH, D.H., and J.H. SMITH, "Measurement of the relativistic time dilation using υ-mesons," *Amer. J. Phys.*, vol. 31, pp. 342-355 (1963).

GALILEI, GALILEO, *Dialogo sopra i due massimi Sistemi del Mondo*, Florence, 1630; Engl. Transl., *Dialogue Concerning the Two Chief World Systems*, Univ. of Calif. Press, Berkeley, 1967.

GERBER, P., "Die räumliche und zeitliche Ausbreitung der Gravitation," *Zeitschr. f. Mathem. u. Phys.*, vol. 43, pp.93-104 (1898); "Die Fortpflanzungsgeschwindigkeit der Gravitation," reprint of 1902 report (see text, p. 171, *Annalen der Physik*, vol. 52, pp.415-444 (1917).

GROVE, D.J., AND J.G. FOX, "*e/m* for 385-MeV protons," *Phys. Rev.*, vol. 90, p. 378 (1953).

HANDBOOK OF MATHEMATICAL FUNCTIONS, National Bureau of Standards, Government Printing Ofice, Washington, DC 1964; reprinted by Dover Publications, New York.

HAZLETT, R., and D. TURNER, *The Einstein Myth and the Ives Papers*, Devin-Adair, Greenwich, Conn., 1979.

HOEK, M., "Détermination de la vitesse avec lacquelle est entrainée une onde lumineuse traversant un mileu en mouvement," *Arch. Néerland.*, Sci. exactes naturelles, vol 3., pp. 180-185, (1868)

IVES, H., "Derivation of the Mass-Energy Relation," *J. Opt. Soc. Am.*, vol. 42, no. 8, pp.182-185, August 1952.

IVES, H., and G.R. STILWELL, "An experimental study of the rate of a moving Atomic clock," *J. Opt. Soc. Amer.*, vol. 28, pp.215-226 (1938); "An experimental study of the rate of an atomic clock, II," *J. Opt. Soc. Amer.*, vol. 31, pp. 369-374 (1941).

KAUFMANN, W., "Die magnetische und elektrische Ablenkbarkeit der Becquerelstrahlen und die scheinbare Masse der Elektronen," *Göttinger Nachr., Mathem.-phys. Kl.*, p.143ff. (1901).

KIRCHNER, F., "Über die Bestimmung der spezifischen Ladung des Elektrons aus Geschwindigkeitsmessungen," *Ann. d. Phys.*, vol. 8, pp. 975-1004 (1931)

LANDAU, L.D., and E.M. LIFSHITS, *Theoretical Physics*, vol. 1, *Mechanics*, Moscow 1965 (Engl. Translation: Pergamon Press, Oxford/New York, 1976)

LAPLACE, PIERRE SIMON, MARQUIS DE, *Mécanique céleste*, Paris, 1799-1825; English transl. by N. Bowdich, Boston, 1839; reprinted by Chelsea Publ., New York 1966.

LIVINGSTON, D.M., *The Master of Light*, Scribner, New York 1973.

LONGMIRE, C.L., *Elementary Plasma Physics*, Wiley-Interscience, New York, 1963

LORENTZ, H.A., "Electromagnetic phenomena in a system moving with any velocity less than that of light," *Proc. Acad. Sci. Amsterdam*, vol. 6, 1904; reprinted in *The Principle of Relativity*, Dover Publ., New York, 1952. See also *The Theory of Electrons*, 2nd ed. 1915, reprinted by Dover Publ., New York, 1952.

MACARTHUR, D.W., "Special relativity: Understanding experimental tests and formulations," *Phys. Rev. A*, vol. 33. no.1, pp.1-5, Jan. 1986.

MACARTHUR, D.W., K.D. BUTTERFIELD, D.A. CLARK, J.B. DONAHUE, P.A.M. GRAM, H.C. BRYANT, C.J. HARVEY, W.W. SMITH, and G. COMTET, "Test of Special Relativity Doppler Formula at $\beta=0.84$," *Phys. Rev. Lett.*, vol. 56, p. 282-285, 27 Jan. 1986.

MANDELBERG, H.I., and L. WITTEN, "Experimental verification of the relativistic Doppler effect," *J. Opt. Soc. Amer.*, vol. 52, pp. 529-536 (1962).

MAJORANA, Q., "Démonstration expérimentale de la constance de vitesse de la lumière réfléchie par un miroir en mouvement," *Compt. Rend. Acad. Franç.*, vol. 165, p.424-426 (1917); "Démonstration expérimentale de la lumière émise par une source mobile," *C.R. Acad. Fr.* vol. 167, pp.71-73 (1918a); "On the Second Postulate of the Theory of Relativity: an experimental demonstration," *Phys. Rev.* vol. 11, pp.411-420 (1918b); "Experimental demonstration of the constancy of the velocity of light emitted by a moving source," *Phil. Mag.* vol.37, p.145-150 (1919).

MICHELSON, A. A., "Relative Motion of Earth and Ether," *Phil. Mag.*, vol. 8, pp.716-719 (1904); "Effect of reflection from a moving mirror on the velocity of light," *Astrophys. J.* vol. 37, pp.190-193 (1913).

MICHELSON, A.A., and H.G.GALE, "The effect of the earth's rotation on the velocity of light," *Astrophys. J.*, vol. 61, pp.137-145 (1925).

NIETO, M. N., *The Titius-Bode Law of Planetary Distances*, Pergamon, Oxford-New York, 1972.

PAGE, LEIGH, "A Derivation of the Fundamental Relations of Electrodynamics from Those of Electrostatics," *Amer. J. Sci.*, Vol. 34, pp.57-68 (1912).

PAGE, LEIGH, and N.I. ADAMS, JR., *Electrodynamics*, Van Nostrand, New York, 1940; reprinted by Dover Publications, New York, 1965.

PENFIELD, P., JR., and H.A. HAUS, *Electrodynamics of Moving Media*, MIT Press, Cambridge, 1967.

POINCARÉ, H., *Archives néerlandaises des sciences exactes et naturelles*, vol. 5, pp. 252-278 (1900); "L'état actuel et l'avenir de la physique mathematique," *La Revue des Idées*, vol. 80, 15 Nov. 1904. Both papers are summarized in [Ives 1952].

SOMMERFELD, A., *Electrodynamics.* Lectures in Theoretical Physics, vol. 3, (1948), based on lectures in 1933-34, Engl Transl., Academic Press, New York, 1964; referred to in text as [Sommerfeld 1933/64], pages refer to English transl., paperback edition.

STRATTON, J.A., *Electromagnetic Theory*, McGraw-Hill, New York 1941.

TAMM, I. E., *Osnovy teorii elektrichestva* (Foundations of the theory of electricity), Gostekhizdat, Moscow 1954.

TOLMAN, R.C., "Some emission theories of light," *Phys. Rev.*, vol. 35, pp.136-143 (1912).

TOMASCHEK, R., "Über das Verhalten des Lichtes ausserirdischer Lichtquellen," *Ann. d. Phys.*, vol. 73, pp.105-126 (1924).

TONNELAT, M.A., *Les Principes de la théorie électromagnétique et de la relativité*, Masson, Paris 1959.

TROUTON, F.T. AND H.R. NOBLE, "The forces acting on a charged condenser moving through space," *Proc. Royal Soc. London*, vol. 72, pp. 132-133, 11 June 1903. This is an abstract of the full paper by both authors, "The mechanical forces acting on a charged condenser moving through space," *Phil. Trans. Royal Soc. London*, ser. A, vol. 202, pp. 165ff., Jan. 1904.

WALDRON, R.A., *The Wave and Ballistic Theories of Light — A Critical Review*, Frederick Muller, London, 1977.

WALZ, D., and R.L. CAREZANI, Calorimetric test of special relativity," *Phys. Rev. A*, vol. 29, no.1, pp. 2110-2113, Apr. 1984.

WHITTAKER, SIR EDMUND, *History of the Theories of Aether and Electricity*, Thomas Nelson & Sons, Edinburgh, 1910 (revised and enlarged 1951, reprinted 1962).

ZAHN, C.T., AND A.H. SPEES, "The specific charge of disintegration electrons from radium E," *Phys. Rev.* vol. 53, pp.365-373 (1938)

ZRELOV, V.P., A.A. TIAPKIN, AND P.S. FARAGO, "Measurement of the mass of 660 MeV protons," *Zhurnal experim. i teoret. fiziki* (Moscow), vol. 34, pp. 555-558, March 1958; English translation in *Soviet Physics-JETP*, vol. 34 (7), no. 3, pp. 384-387, Sept. 1958.

INDEX

THE AUTHOR

was born in Prague, Czechoslovakia, in 1924. He received his M.Sc. and Ph.D in Electrical Engineering from Prague Technical University in 1942 and 1955, and a Dr.Sc. degree from the Czechoslovak Academy of Sciences in 1962. In 1963 he was invited by the University of Colorado as a Visiting Professor, defected to the US, became a US citizen, and taught at the University of Colorado until his retirement in 1981.

He has published more than 60 scientific papers, mostly devoted to electromagnetics and probability theory. His scientific books include *The Scattering of Electromagnetic Waves From Rough Surfaces* (With A. Spizzichino, 1963, reprinted 1987), *The Depolarization of Electromagnetic Waves* (1968), and *The Structure of Language* (1973); his textbooks include *Probability in Communication Engineering* (1967), *Elementary Queuing Theory and Telephone Traffic* (1967), and *Orthogonal Polynomials for Engineers and Physicists*. His best selling popular books are *A History of π (Pi)* (1970, 5th ed. 1981) and *The Health Hazards of NOT Going Nuclear* (1976, 10th printg. 1985).

He is editor and publisher of the "pro-science, pro-technology, pro-free enterpise" monthly newsletter *Access to Energy*, now in its 15th year with some 7,000 readers.